SpringerBriefs in Education

We are delighted to announce SpringerBriefs in Education, an innovative product type that combines elements of both journals and books. Briefs present concise summaries of cutting-edge research and practical applications in education. Featuring compact volumes of 50 to 125 pages, the SpringerBriefs in Education allow authors to present their ideas and readers to absorb them with a minimal time investment. Briefs are published as part of Springer's eBook Collection. In addition, Briefs are available for individual print and electronic purchase.

SpringerBriefs in Education cover a broad range of educational fields such as: Science Education, Higher Education, Educational Psychology, Assessment & Evaluation, Language Education, Mathematics Education, Educational Technology, Medical Education and Educational Policy.

SpringerBriefs typically offer an outlet for:

- An introduction to a (sub)field in education summarizing and giving an overview of theories, issues, core concepts and/or key literature in a particular field
- A timely report of state-of-the art analytical techniques and instruments in the field of educational research
- A presentation of core educational concepts
- An overview of a testing and evaluation method
- A snapshot of a hot or emerging topic or policy change
- An in-depth case study
- A literature review
- A report/review study of a survey
- An elaborated thesis

Both solicited and unsolicited manuscripts are considered for publication in the SpringerBriefs in Education series. Potential authors are warmly invited to complete and submit the Briefs Author Proposal form. All projects will be submitted to editorial review by editorial advisors.

SpringerBriefs are characterized by expedited production schedules with the aim for publication 8 to 12 weeks after acceptance and fast, global electronic dissemination through our online platform SpringerLink. The standard concise author contracts guarantee that:

- an individual ISBN is assigned to each manuscript
- each manuscript is copyrighted in the name of the author
- the author retains the right to post the pre-publication version on his/her website or that of his/her institution

Aaron M. Ellison · Manisha V. Patel

Success in Mentoring Your Student Researchers

Moving STEMM Forward

 Springer

Aaron M. Ellison 🆔
Sound Solutions for Sustainable Science
LLC
Boston, MA, USA

Manisha V. Patel
Sound Solutions for Sustainable Science
LLC
Boston, MA, USA

ISSN 2211-1921   ISSN 2211-193X (electronic)
SpringerBriefs in Education
ISBN 978-3-031-06644-3   ISBN 978-3-031-06645-0 (eBook)
https://doi.org/10.1007/978-3-031-06645-0

This Springer imprint is published by the registered company Springer Nature Switzerland AG
The registered company address is: Gewerbestrasse 11, 6330 Cham, Switzerland

Preface

Research is at the core of scientific inquiry. Mentoring students in research as early as possible in the course of their education gives them opportunities to critically examine and, ideally, affirm their commitment to a career in Science, Technology, Engineering, Mathematics, or Medicine (STEMM); contributes to diversification of the STEMM workforce; and helps build networks of the next generation of scientists and STEMM educators. Indeed, many career scientists credit their participation in undergraduate research experiences with cementing their interest in STEMM fields and launching them onto successful career trajectories in academics, government agencies, nonprofit organizations, and the private sector.

Mentoring undergraduate researchers has additional rewards. More senior researchers educate nascent scientists and create more inclusive STEMM disciplines while advancing their own science in partnership with smart, enthusiastic, motivated, and hard-working students. Junior mentors, including graduate students, have opportunities to develop professional skills in advising and teaching. Both these forms of teaching and mentoring measure success not by grades but by enhancements in understanding of complex topics and critical thinking skills.

Each mentor has a different approach and uses different methods in working with their students, but there are common approaches and practices that mentors can use to contribute to the success of every student researcher. In this book, *Success in Mentoring Your Student Researchers: Moving STEMM Forward*, and its companion, *Success in Navigating Your Student Research Experience: Moving Forward in STEMM*, we draw on our combined more than 50 years of experience and work as undergraduate researchers, research mentors, program managers, and undergraduate research program leaders to identify these common approaches. This book brings together in one place a set of best practices that mentors can use to engage with and succeed in intensive, experiential research with students outside a classroom. The overall goal of this book is to guide STEMM researchers toward a deeper understanding of how to be a mentor to student researchers and to commit to mentorship that builds diverse and inclusive communities of collaborative researchers.

The mentored undergraduate research experiences we focus on in this book normally are full-time, usually paid, summer, semester, or year-long positions such as "REU" positions (in the USA) or "internships" (worldwide), but may also be senior thesis projects or a semester- or year-long independent study "classes" for credit. Mentoring a student through intensive experiential research differs substantially from classroom-based teaching, even when the latter uses the variety of well-established modes and mechanisms of active learning, such as inquiry- and project-based learning-by-doing, flipped classrooms, and place-based study.

Actually mentoring undergraduate research is a multifaceted process that involves understanding the essential elements of mentorship, and funding and recruiting mentees; mentoring the research itself; and finally evaluating oneself and one's mentees, possibly continuing research with them, and certainly fledging them into the broader world of STEMM professionals and careers. These three activities are mirrored in the three parts of the book, each of which is divided into three chapters. Most of the material in each chapter is on how individuals can be better mentors for their mentees, but the last section of each chapter expands our vision of mentorship to include how one can become a mentor of mentors and develop and lead undergraduate research programs. Most chapters also have one or more Text Boxes and Vignettes. The Text Boxes concisely illustrate key ideas or tools, or suggest a set of questions or prompts to guide researchers as they develop and hone their skills in mentoring. The Vignettes are short, first-person narratives solicited through an open call on social media to current and former mentors about their own experiences mentoring student researchers.

In Chap. 1 of Part I, we start by introducing our definition of mentors and mentorship in STEMM:

Mentors see potential in their mentees and empower them to become self-aware, independent-minded scientists and scientifically literate members of their communities.

Through an emphasis on co-learning and empowerment, we distinguish mentorship from advising, supervising, and managing students. We recognize that mentoring requires more time, energy, and commitment than advising, supervising, or managing, and we provide tools for prospective and current mentors to use to work through their goals and expectations for mentorship. We then give pointers to the range of available resources to continue to improve skills in mentorship.

STEMM research at any level is not free. Chapter 2 discusses how to financially support your undergraduate mentees, drawing on the billions of dollars that are spent worldwide by government agencies, private foundations, individual institutions, and companies and industries in the private sector on undergraduate research experiences. Chapter 3 concludes this first part by highlighting the importance of recruiting, hiring, and mentoring a diverse cadre of undergraduate researchers and emphasizes the importance of "recruiting for potential."

In Part II, we dive into the details of how mentoring undergraduate research differs substantially from teaching them the elements of research in their college classes, even those where active learning is a central component. We consider the undergraduate research experience itself as a "three-legged stool," whose legs are research, education, and community. The balance among these legs helps to ensure success in research and careers in STEMM.

Although doing research itself is the *raison d'être* of a mentored research experience, if a mentee does not feel like a part of a team or community of like-minded peers and role models, even the most engaging research project will fail to capture their imagination. Thus, this part begins with a chapter on how to build, organize, and manage an inclusive community of researchers (Chap. 4). We then move onto the essentials of mentoring a student in doing research that complements the mentor's own. This chapter emphasizes three points: (1) the importance of setting expectations (for both mentors and mentees); (2) empowering mentees through feedback, encouragement, and allowing them to "take ownership" of a part of a larger research endeavor while preserving its conceptual integrity; and (3) working with mentees to help them identify opportunities for further intellectual and career growth in STEMM.

The last chapter in Part II outlines ways for mentors to effectively teach a range of professional skills that all scientists need to be successful in STEMM. These skills include writing research proposals, navigating permitting requirements, communicating results to a range of audiences, identifying career trajectories, and building professional networks (Chap. 6). It is far better to develop these skills through training sessions and participatory workshops than by expecting one's mentees to just "pick them up" when they are needed.

Finally, mentorship does not stop at the end of the summer, term, year, or undergraduate research program. Thus, the three chapters in Part III examine mentorship beyond a single mentored research experience. In Chap. 7, we present various ways to evaluate the student and their research, as well as the mentor's activities. We emphasize the importance of grounding evaluation in theories of learning to allow for generalizations and inferences across mentored undergraduate research experiences and programs. In Chap. 8, we outline ways to continue to support students to continue their research beyond the formal time frame of their individual research experience. We conclude the book with a discussion of fledging one's mentees and continuing the mentor-mentee relationship in the long-term (Chap. 9).

Like doing research itself, writing and publishing a book is a team effort. Our ideas about mentored undergraduate research have been refined by our work with hundreds of students and dozens of our colleagues—including administrative and support staff, graduate students and post-docs, senior researchers, and faculty members—at the institutions where we have worked (Swarthmore College, Mount Holyoke College, Rutgers University, the University of Vermont, and Harvard University) and with whom we collaborate through various professional networks (BIO-REU, ESA-SEEDS, LTER, OBFS, and UFERN). We are deeply grateful to Beth Fischer (Assistant Professor in the School of Education at the University of Pittsburgh) and Michael Zigmond (Professor Emeritus of Neurology, Psychiatry, and Neurobiology at the University of Pittsburgh) for teaching us the importance of "survival skills" in STEMM through their workshops and inviting Aaron Ellison to participate in their "trainer-of-trainers" conferences in the 1990s; and to Brad Rose (Brad Rose Consulting) and Andrew McDevitt (Lecturer in Biology at the University of Colorado, Denver) for teaching us how to evaluate undergraduate research programs and then working with us and undergraduate researcher Relena Ribbons (now Assistant Professor of Geosciences at Lawrence University) to evaluate the Harvard Forest

Summer Research Program in Ecology. Last, we thank Claudia Acuna, our editor at Springer Nature for taking on this book and seeing it through to publication; two anonymous reviewers who provided useful comments on the book proposal and the final manuscript; and Rajan Muthu and his production team in Chennai who wrangled our manuscript into its final form.

Over the last three decades, our work as undergraduate research mentors and undergraduate research program co-directors has been supported not only by our home institutions but also by a range of competitive awards. For support of individual undergraduate researchers, these have included grants from the DOE (award no. DE-FG02-08ER64510); the Ellen P. Reese Fund at Mount Holyoke College; the HHMI; the Massachusetts Natural Heritage and Endangered Species Program; The Mellon Foundation; the US National Institute of Climate Change Research (NICCR); NSF (award nos. BSR 9107915; DEB 9253743, 9741904, 9805722, 9942207, 0115145, 031361, 0235128, 0422750, 0520792, 0528625, 0541680, 0722588, 0816508, 0902592, 1025362, 1110434, 1144056, 1136646, 1518653); and the Orchards Golf Course in South Hadley, Massachusetts. For support of undergraduate research programs, these have included grants from the HHMI (award no. 71196-505002); the NSF (award nos. DBI 0330605, 0422745, 0452254, 0520794, 0618448, 0812997, 0930516, 1003938, 1111158, 1224437, 1239937, 1341122, 1446653, 1459519, 1535283; ACI 1450277); NASA (NNX10AT52A); and the Sherman Fairchild Foundation.

Finally, we especially acknowledge the influence and impact of two permanent program directors at NSF—Sally O'Connor and Sonia Ortega—whose complementary visions for undergraduate research have shaped our own and who have been instrumental in advancing undergraduate research programs across the US and around the world.

Boston and Greenfield, USA   Aaron M. Ellison
and Singapore   Manisha V. Patel
January 2021

Contents

Part I   Preparing to (be a) Mentor

1   Understanding Mentorship 3
  1.1   Mentors and Mentorship Defined 4
     1.1.1   The Essence of Mentors and Mentorship
             in Undergraduate Research 5
     1.1.2   Mentorship is Not Supervising or Advising 5
  1.2   Why Mentor? 6
  1.3   Group Mentorship: Because You Shouldn't Have to Mentor
        Alone 7
  1.4   Becoming a Mentor 8
  1.5   From Mentors to Program Directors: Becoming a Mentor
        of Mentors 9
  1.6   Take-Home Messages 9
  References 9

2   Funding Undergraduate Research 11
  2.1   Which Comes First: Defining the Research, Identifying
        a Student, or Finding the Funding? 12
  2.2   The Costs of Undergraduate Research 13
     2.2.1   Stipend or Salary, Credit or Volunteer? 13
     2.2.2   Subsistence Costs 13
     2.2.3   Travel Costs 14
     2.2.4   Other Direct Costs 15
     2.2.5   Administrative and Other Hidden Costs 15
  2.3   Defining Mentored Research Projects: How Much Student
        Independence Can You Afford? 15
     2.3.1   From First-Time Researchers to Self-Aware
             and Independent Mentees 16
     2.3.2   Can You Afford Independent Students? 16
  2.4   Finding Funding for Mentored Student-Research Projects 17
     2.4.1   Finding Funding for Mentor-Defined Projects 17

          2.4.2   Finding Funding for Independent Mentees   17
    2.5   From Mentors to Program Directors: Funding a Whole Program   19
          2.5.1   It Really Does Take a Village   19
          2.5.2   Supporting the Village   20
          2.5.3   Ensuring Equity   20
    2.6   Take-Home Messages   21
    References   21

3   **Recruiting and Selecting Students**   23
    3.1   Setting Priorities   24
    3.2   Recruitment   25
          3.2.1   Why Not Hand-Pick a Rock Star?   26
          3.2.2   The Value of a Formal Recruitment Process   27
          3.2.3   The Elements of a Formal Recruitment Process   27
    3.3   Reviewing Applications   29
    3.4   Interviewing Applicants   30
          3.4.1   Set the Tone   30
          3.4.2   Platform and Setting   30
          3.4.3   Consistency in Questions   31
          3.4.4   Wrapping up the Interview   31
    3.5   Selecting and Hiring Students   32
    3.6   From Mentors to Program Directors: Recruiting New Mentors   32
          3.6.1   Research Mentors   33
          3.6.2   Trainers and Facilitators   33
          3.6.3   Support Staff to Support Diverse Students   33
          3.6.4   Ensuring Equity   34
    3.7   Take-Home Messages   34
    References   34

Part II   **Mentoring Student Researchers**

4   **Building a Research Community**   37
    4.1   STEMM Communities   38
    4.2   Building a Collaborative Research Team   39
          4.2.1   Roles in a Professional Community are Not Static   40
          4.2.2   Bringing New People into Your Research Team   41
          4.2.3   The Importance of Formal and Informal Communication ...   42
    4.3   From Mentors to Program Directors: Developing Communities
          of Researchers   43
    4.4   Take-Home Messages   44
    References   44

5 Doing Research with Undergraduates 47
  5.1 Setting Expectations 48
      5.1.1 The Research Comes First, But it is First Among Equals 48
      5.1.2 The Importance of Communication 49
  5.2 Empowering Your Students 50
      5.2.1 Ceding Some Intellectual Ownership 50
      5.2.2 The Importance of Regular Feedback 51
  5.3 Future Research with and by Your Mentees 52
  5.4 From Mentors to Program Directors: The Science Behind
       Successful Undergraduate Research Programs 52
  5.5 Take-Home Messages 53

6 STEMM Education is More Than Training 55
  6.1 Basic Training 56
      6.1.1 Ethical and Responsible Conduct of Research 56
      6.1.2 Doing Research Safely 57
      6.1.3 Teaching Technical Skills 58
  6.2 Teaching the Research Process 58
      6.2.1 Experience Matters 59
      6.2.2 Teaching Without Giving Grades 59
  6.3 Professional Development 60
  6.4 From Mentors to Program Directors: Teaching Skills Needed
       by STEMM Professionals 61
  6.5 Take-Home Messages 62
  References 62

Part III Mentoring Beyond the Research Experience

7 Evaluation 65
  7.1 Why Evaluate? 66
  7.2 Evaluation in Practice, in Theory, and in Practice Grounded
       in Theory 66
      7.2.1 Are You Evaluating for Yourself or For Others? 67
      7.2.2 The Value of Individual Case Studies 67
      7.2.3 National-Level, Cross-Program Surveys 68
      7.2.4 The Importance of Theory 69
  7.3 Theories and Resources for Evaluating Undergraduate Research
       Programs 69
      7.3.1 Educational Theories for Evaluation 69
      7.3.2 Partnering for Success 70
  7.4 Learning from Evaluations and Evolving Undergraduate
       Research Experiences 71
  7.5 From Program Directors to Mentors: Closing the Feedback Loop ... 71
  7.6 Take-Home Messages 72
  References 72

8   Continuing the Research Experience 75
  8.1   How Did It Go? 76
        8.1.1   Take Time to Reflect 76
        8.1.2   Evaluate Yourself and Your Team 77
        8.1.3   Provide Feedback to Your Mentees 77
  8.2   Finishing up Your Mentees' Research Experience 77
        8.2.1   Tying up Loose Ends 78
        8.2.2   Data Management and Data Archiving 78
        8.2.3   Presentations at Meetings 79
        8.2.4   Writing It up 79
  8.3   Additional Research Opportunities with and for Your Mentees 80
        8.3.1   More Undergraduate Research with You 80
        8.3.2   Undergraduate Research with Others at Your Own
                Institution or Elsewhere 81
  8.4   What's Next for You? 82
  8.5   From Mentors to Program Directors: Facilitating and Sustaining
        Further Undergraduate Research 82
        8.5.1   Facilitating Interactions 82
        8.5.2   Incentives for Continuing Undergraduate Research 83
        8.5.3   Take Advantage of Online Platforms 83
  8.6   Take-Home Messages 85
  References 85

9   Fledging Your Mentees 87
  9.1   Post graduate Directions 87
        9.1.1   Employment 88
        9.1.2   Post-baccalaureate Programs 88
        9.1.3   Graduate and Professional Schools 89
  9.2   Letters of Recommendation 90
  9.3   From Mentees to Mentors 90
  9.4   From Mentors to Program Directors: Finding Your Successors 91
  9.5   Take-Home Messages 92
  References 92

Notes 93

Acronyms

CHAT   Cultural Historical Activity Theory, a general theory that describes how learning is culturally mediated.

CIMER   The Center for the Improvement of Mentored Experiences in Research, an organization that has developed a number of platforms for assessing outcomes of undergraduate research.

CRediT   Contributor Roles Taxonomy is a system that provides clear guidelines for maintaining integrity and transparency in determining and asserting (co-)authorship of research papers.

FAIR   Findable, Accessible, and Interoperable and Reusable. FAIR refers to a set of guidelines that encapsulate current best practices for responsible and ethical data management.

GPA   Grade-point Average.

HHMI   The USA-based Howard Hughes Medical Institute.

IACUC   Institutional Animal Care and Use Committee. These institution-based committees ensure that research involving animals (usually only vertebrate animals) is conducted ethically and safely, minimizes harm to the animals involved, and conforms to all applicable laws and regulations.

IAT   Implicit Association Test. One of the most widely used and best validated instruments to assess an individual's implicit bias.

IRB   Institutional Review Board. These institution-based committees ensure that research involving humans is conducted ethically and safely, minimizes harm to the people involved, and conforms to all applicable laws and regulations.

NIH   US National Institutes of Health.

NSF   US National Science Foundation.

OBFS   Organization of Biological Field Stations.

PUI   Primarily Undergraduate Institutions, a categorization used by the NSF to describe accredited colleges and universities (including two-year community colleges) that award Associate's degrees, Bachelor's degrees, or Master's degrees in NSF-supported fields, but have awarded 20 or

	fewer PhD/DSc degrees in all NSF-supported fields during the combined previous two academic years.
RECCS	Research Experience for Community College Students, a cross-directorate program of the NSF.
REU	Research Experience for Undergraduates, a cross-directorate program of the NSF.
SALG	The Student Assessment of Learning Gains, a survey instrument that is commonly used to evaluate the outcomes of undergraduate research experiences.
STEM	Science, Technology, Engineering, or Mathematics.
STEMM	Science, Technology, Engineering, Mathematics, or Medicine.
SURE	Survey of Undergraduate Research Experiences. SURE was one of the first broad-based assessments of undergraduate STEMM research. It has been superseded by URSSA and instruments developed by CIMER.
URSSA	The Undergraduate Research Student Self-Assessment survey.

List of Boxes and Vignettes

Box 1.1   Self-reflection: Who are and were your mentors? 6
Box 1.2   Resources for mentors 8
Vignette 2.1 Leveraging undergraduate research to launch a research
             program 18
Box 3.1   Implicit bias: recognize it and reduce it 25
Vignette 3.1 Successful outcomes from recruiting for potential 26
Box 4.1   The twelve most important elements of successful
             research teams 40
Vignette 4.1 Building diverse teams 41
Vignette 5.1 This time it didn't work: remote mentorship and the
             magnetic field of a tractor 51
Vignette 6.1 Formal training helps you and your students mitigate bad
             outcomes 57
Box 8.1   Self-reflection: Did you accomplish your goals as a
             mentor? 76
Vignette 8.1 Sustaining undergraduate research programs by adapting
             to the times 84
Vignette 9.1 Long-term rewards of mentoring 90

Part I
Preparing to (be a) Mentor

In Part I of this book, we start out by defining mentors and mentorship in STEMM. Through an emphasis on co-learning and empowerment, we distinguish mentorship from advising, supervising, and managing students. Mentoring undergraduate researchers requires personal commitment, but STEMM research is not free. The second chapter discusses how to financially support your undergraduate mentees. We conclude this section by highlighting the importance of recruiting, hiring, and mentoring a diverse cadre of undergraduate researchers and emphasize the importance of "recruiting for potential."

Chapter 1
Understanding Mentorship

Abstract Our focus in this book is on mentors and mentoring, not supervising, managing, or advising. This chapter begins by defining mentors and mentorship in the broad sense and then, more specifically, in the context of working with undergraduate researchers; this contextual definition of mentorship is used throughout this book. The last section of this chapter—like all subsequent chapters in this book—expands a vision of mentorship to include becoming a mentor of mentors and developing and leading undergraduate research programs. The goal of this chapter, and of the entire book, is to help the reader achieve a deeper understanding of how to be a mentor to student researchers and to commit to mentorship that builds diverse and inclusive communities of collaborative researchers.

All of us who work with students play many roles. At different times, we may be supervisors, managers, advisors, or mentors. Like an actor covering multiple characters in a single play, when we are in one of these positions we adopt its unique characteristics and others that it shares with one or more of the other roles (Fig. 1.1). Our focus in this book is on mentors and mentoring, not supervising, managing, or advising. To ensure we're all acting in the same play, we begin by defining mentors and mentorship in the broad sense and then, more specifically, in the context of working with undergraduate researchers. We will use this latter contextual definition throughout this book.

This first chapter includes two Text Boxes. Box 1.1 prompts you to reflect on your own mentors and your role as a mentor. Box 1.2 provides pointers to additional resources where you can learn, develop, and refine your mentorship skills.

Finally, the last section of this chapter—and all subsequent chapters in this book— broadens our visions of mentorship to include becoming a mentor of mentors and developing and leading undergraduate research programs. We hope you come away from this chapter ready to dive into the rest of this book, achieve a deeper understanding of how to be a mentor to your students, and commit to mentorship that builds diverse and inclusive communities of collaborative researchers.

© The Author(s), under exclusive license to Springer Nature Switzerland AG 2022   3
A. M. Ellison and M. V. Patel, *Success in Mentoring Your Student Researchers*,
SpringerBriefs in Education,
https://doi.org/10.1007/978-3-031-06645-0_1

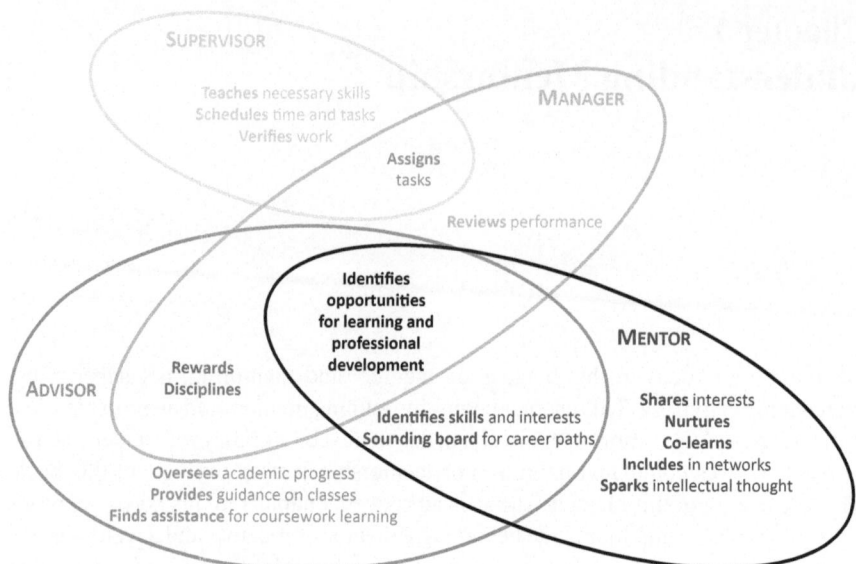

Fig. 1.1 The overlapping roles of supervisors, managers, advisors, and mentors

1.1 Mentors and Mentorship Defined

There are dozens of definitions of mentors and mentorship in internal presentations, how-to manuals, reviews published in journal articles, and books that span disciplines ranging from education and organizational psychology to every field considered a part of Science, Technology, Engineering, Mathematics, or Medicine (STEMM) (National Academies of Sciences, Engineering, and Medicine, 2019). The broadest dictionary definition of a mentor is "a person who acts as guide and adviser to another person, especially one who is younger and less experienced" (OED, 2020). In colleges and universities, a mentor is usually an experienced, trusted person—often a professor, senior staff member, or fellow student—who offers guidance, support, and counsel to someone who is less experienced and interested in collaboratively learning about an area of shared interest. We emphasize that your main goal as a mentor should be to enhance careers of others, not to advance your own career.

STEMM theorists and practitioners, including Einstein and Edison, have argued that the association between a mentor and a mentee is the most significant collaborative relationship in one's career (Ma et al., 2020). The National Academy of Sciences (2019, p. x) calls mentorship a series of "high-stakes, interpersonal encounters and exchanges" that acts as a "catalytic factor to unleash individuals' potential for discovery, curiosity, and participation in STEMM" (p. xi). Whereas every academic relationship involves teaching and learning (Fig. 1.1), not all develop into mentor-mentee relationships. Rather, you will find that becoming a mentor takes time, trust, and mutual respect.

1.1.1   The Essence of Mentors and Mentorship in Undergraduate Research

Our mantra as undergraduate-research mentors and program leaders is that:

> Mentors see potential in their mentees and empower them to become self-aware, independent-minded scientists and scientifically literate members of their communities.

For mentors and mentees, reciprocal and reciprocated trust and communication are key components of their relationship. As in any interpersonal relationship, you and your mentees need to continually nurture, build, and re-create your ways of working together effectively and productively. It's also worth regularly reminding yourself that mentorship will not always be smooth sailing. If your relationship with your mentee runs into rough water or falls apart, it's not because you're a failure or your mentee cannot be a scientist. Rather, challenges in your relationship present new opportunities to learn how to work with each of your students differently and what behaviors to avoid or change.

1.1.2   Mentorship is Not Supervising or Advising

There are many ways that you as a more senior scientist—whether you're a long-serving professor or a new graduate student or post-doc—interact with undergraduates, and not all of these interactions involve mentorship (Fig. 1.1). For example, your routine tasks of academic advising—common examples of which include checking course and degree requirements, approving schedules, negotiating extensions of deadlines with colleagues or deans—rarely empower independence or engender self-awareness. Similarly, you are unlikely to be mentoring when you're supervising student technicians or research assistants, approving their time sheets, setting a work schedule, checking that the work is completed, and correcting any errors.

Although research advisors are often referred to as mentors, as a research advisor for an undergraduate (or graduate) student you may or may not become a mentor. Because a defining feature of mentorship is the reciprocity of the mentor-mentee relationship, you are less likely to be a mentor if you hand a student a research project but then leave them to sink or swim with it than if you work closely with them through the whole process from project ideation and design to its completion. Even then, imbalances in (perceived) power, responsibilities, or other inequities, or your overlapping roles as an employer, supervisor, and research advisor can leave your student feeling poorly supported in their goals as a nascent researcher while you wonder why you're not getting through to them.

Finally, becoming and being a mentor takes more than reading a myriad of mentorship manuals (including this one!) or attending mentoring workshops and training courses. These are all helpful, but being a mentor to your students while attending to all other aspects of your job, family, and daily life takes time, energy, and serious commitment. In our own experience, it is too much to expect any scientist to be a true mentor to every one of their students, and it is unrealistic for any student to expect every more-senior scientist they interact with to become their mentor *du jour*. Before continuing on to the next section, take some time to reflect on your own experience with and as mentors (Box 1.1).

Box 1.1   Self-reflection: Who are and were your mentors?

Stop reading for a few minutes and reflect on who your own mentors were or are, and why you consider them to be your mentors. Ask yourself:

- Who were and are my mentors?
- What interests did/do I share with them?
- How did each of them influence my development (both as a scientist and a person)?
- What barriers did they help me overcome?
- What did I teach (give back to) them?

Now, consider the interactions you have had or are having now with your students and colleagues. Ask yourself:

- How am I influencing their development as scientists and peers?
- What barriers am I helping them to overcome? If I can't help, am I able to connect them with others of different genders, race or ethnicity, abilities, or economic backgrounds who can provide additional support or guidance?
- What am I learning from them that's making me a better scientist and person?
- How have my experiences as a mentor changed my perception of mentorship?

1.2   Why Mentor?

Scientists work in teams to address complex questions, test models and hypotheses, or contribute to solutions of socially relevant problems. Our teams may include peers and colleagues, post-docs, graduate students, undergraduates, high-school interns, or volunteers. Hiring an undergraduate student to work as an assistant in your lab or office or to do field work is cost-effective and efficient, and may require little training and supervision. Indeed, you may even delegate hiring and supervising undergraduates to graduate students or post-docs. But directly or indirectly supervising student workers is not mentorship. Because you will put in much more effort and time commitment as a mentor than as an advisor to your students or a supervisor and manager of your technicians, you would hope that there would be added value of mentoring that compensates for its inefficiency and mental, physical, or even financial costs (the last of these is the focus of Chap. 2).

The benefits of mentoring start with the satisfaction that comes from doing your job well. Part of your job as a more experienced scientist is to share your love of science with up-and-coming scientists. Many of your undergraduate students will not yet have committed to a career in STEMM, will be unsure of their abilities, and are likely to be struggling with a range of other issues and exigencies—many completely unrelated to science courses or research—as they explore and define their identities. These may be irrelevant to you as a supervisor or faculty advisor, but as a mentor, you can make a real difference in your students' lives by listening, understanding, and supporting them as they navigate their own paths to becoming scientists.

Mentoring also can change the way you teach. Rather than lecturing *to* your students or flipping the classroom so that your students teach themselves (and perhaps some of their peers), bringing principles of mentoring into your classroom sets you and your students up as co-teachers and co-learners. Although you may cover less material in a course where you are more of a mentor than an instructor, both you and your students will internalize more of it and come away with a deeper appreciation of the process of learning.

Mentoring also can refine or redirect your own research. You may be pleasantly surprised to find that your undergraduate mentees—given the opportunity to contribute as full participants in your research group—change the way you think about a question or come up with the key insight that makes for your next successful grant proposal.

Finally, each of us got to where we are because of the guidance and influence of at least one mentor (Box 1.1). To effectively move science and the community of scientists into the future, you want to be the mentor of your own students that your mentor was for you. Pay it forward.

1.3 Group Mentorship: Because You Shouldn't Have to Mentor Alone

You want to be a mentor—and a good one, at that—but for whatever reason, you don't have the time, energy, or skills to "do it all" yourself. You will never have an identical background or exactly the same experiences as your mentees, but others in your research group may share some of what you don't or can't. By diversifying from a classic one mentor-one mentee relationship to include others in your research group as co-mentors, you can add new avenues of support and learning for your mentees. The mentoring "load" also is lightened on everyone, and you'll learn new mentoring skills, too.

The best mentor groups are diverse, as diversity provides a range of perspectives and insights that no one person from one gender, race, class, religion, or circumstance can provide. But diversity alone is not enough (Shavit and Ellison, 2021). As the leader of your research group, and perhaps the mentor of your co-mentors, you need to be fully inclusive in supporting your team and creating the space for them to thrive and succeed.

Group mentorship can take many forms. These range from the informal and organic to the more formal and structured. Informal group mentorship can occur during (ir)regular conversations over morning coffee or afternoon tea where current issues and ideas are shared. Formal group mentorship starts with self-reflection and focused discussion with your colleagues, post-docs, and graduate students that reveals strengths and weaknesses of each member of your research team. Through these conversations, you will discover who has skills or desires for certain parts of mentoring. Through side-by-side practice, some members of your team can facilitate others who want to develop or hone their mentoring skills.

1.4 Becoming a Mentor

We began this chapter by illustrating the different roles we play when working with our students (Fig. 1.1). Although we have all had courses in science and most of us have attended workshops on writing proposals and effective teaching, few of us have had any formal courses in supervision, management, or advising, much less mentoring. (See Ruben, 2020 for an entertaining essay on the effects of this lack of education, and Stelter et al., 2021 for a discussion of what should be included in good mentor training courses.) There are a wide range of resources, including books, discussion groups, workshops, mini-courses that can introduce you to core concepts of mentoring and mentorship (Box 1.2). These are most effectively used after you have committed yourself to becoming a mentor. But at the end of the day, *being* a mentor requires more than reading, course attendance, and certificates. Mentoring is a way of acting, doing, behaving, and living.

Box 1.2 Resources for mentors

A partial list of online guides, training courses, workshops, etc. for STEMM mentors
- NIH Guide to Training and Mentoring
- Oak Ridge Institute for Science and Education
- Pathways to Science
- The Mentoring Institute
- The Science of Effective Mentorship in STEMM
- Women in STEM Mentorship Program

1.5 From Mentors to Program Directors: Becoming a Mentor of Mentors

As a mentor, you may find yourself interested in working with your colleagues to take your ideas of mentorship and day-to-day activities as a mentor to another level. For many scientists, this means developing a more formal undergraduate research program that brings together a diversity of researchers and provides opportunities for more students than can fit in one research group.

There are many good reasons to organize, develop, and run undergraduate research programs. By leading such programs, you can bring new ideas and collaborations to your department, school, or institution. You can make more rapid changes in institutional cultures, enhance student diversity and inclusion initiatives, or create and provide multiple opportunities for your graduate students, post-docs, and colleagues in group mentorship. You will usually find many colleagues who are willing to work with you on these programs. They want to learn new skills, become better advisors and mentors, and have additional opportunities to work with students, but they are also looking to someone else who will organize and lead broader efforts.

The rewards of running an undergraduate research program will be accompanied by changes in your day-to-day working life. You will need to learn new skills and deploy new tools for management and administration. You might view these activities as burdens that take time away from your "real" work (i.e., teaching, lab or field research, publishing, and grant-writing), but you shouldn't. Selfishly, you may find that management and administrative experience add luster to your CV and provide you with a leg up for promotions to department chair, academic dean, provost, or vice-president for research positions. More broadly, improving institutional cultures and having positive effects on your students and colleagues creates lasting change that will be remembered for many years.

1.6 Take-Home Messages

✔ Being a mentor is different from being an advisor, manager, or supervisor.
✔ Mentor-mentee relationships are some of the most significant collaborative relationships in scientists' careers.
✔ A group of diverse mentors is more powerful than a single one.

References

Ma, Y., Mukherjee, S., & Uzzi, B. (2020). Mentorship and protégé success in stem fields. *Proceedings of the National Academy of Sciences, USA, 117*(25), 14077–14083.

National Academies of Sciences, Engineering, and Medicine. (2019). *The science of effective mentorship in STEMM*. Washington, DC: The National Academies Press.

OED. (2020). "mentor", n. OED. https://www.oed.com/view/Entry/116575.

Ruben, A. (2020). Scientists aren't trained to mentor. That's a problem. *Science, 369.*

Shavit, A., & Ellison, A. M. (2021). Diversity is conflated with heterogeneity. *The Journal of Philosophy, 118,* 525–548.

Stelter, R. L., Kupersmidt, J. B., & Sump, K. N. (2021). Establishing effective STEM mentoring relationships through mentor training. *Annals of the New York Academy of Sciences, 1483,* 224–243.

Chapter 2
Funding Undergraduate Research

Abstract This chapter discusses the financial costs of supporting undergraduate researchers and undergraduate research experiences. It is important to identify these costs before committing to mentoring undergraduate researchers and to ensure that they are compensated fairly and equitably. There is a range of external funding sources, including the government, foundations, NGOs, and the private sector, that support undergraduate research. Many institutions also have internal or discretionary funds that are available to students or more senior researchers that can support mentored research.

Doing research is not free, and undergraduate research experiences have obvious and non-so-obvious financial costs. In this chapter we discuss the financial costs of supporting undergraduate researchers and undergraduate research experiences. There is a range of external funding sources, including the government, foundations, NGOs, and the private sector, that support undergraduate research. Many institutions also have internal or discretionary funds that are available to students or more senior researchers that can support mentored research.

As you're reading this chapter, keep in mind that as mentors, we tend to think that these are "our" costs—student stipends, lab or field costs, etc.—but we also need to be aware that our mentees have their own financial obligations. Time is not infinitely elastic, and for many undergraduates, doing research means not doing something else, which might be a job needed to pay their tuition, room, or board. Although your mentees may envision a STEMM career and their research experience with you as the first step toward achieving that goal, right now they also need to pay the bills; the stipends or salaries paid for participating in a research experience could be their primary source of income. You may initially find it uncomfortable to discuss financial issues with your mentees, but it's important to realize that they are independent individuals with real-world responsibilities—just like you.

Finally, we emphasize that mentoring undergraduate researchers also has real benefits and rewards. Although these may not be as financially tangible as paying a student-researcher's stipend, the benefits are large and the rewards long-lasting. The one Vignette in this chapter, Vignette 2.1, highlights some aspects of the benefits and rewards of mentoring undergraduates and how to leverage independent undergraduate

A. M. Ellison and M. V. Patel, *Success in Mentoring Your Student Researchers*,
SpringerBriefs in Education,
https://doi.org/10.1007/978-3-031-06645-0_2

research to launch a long-term research program. These benefits echo throughout the other chapters of this book and in our companion book, *Success in Navigating Your Student Research Experience*.

2.1 Which Comes First: Defining the Research, Identifying a Student, or Finding the Funding?

There are at least two ways you might start thinking about funding undergraduate research and undergraduate researchers. On the one hand, you can start by defining the kind of research you want to do with undergraduates, then determine your expectations for the students and consider how much independence you expect of them, and last, determine the financial costs and how to cover them. On the other hand, you might think first about the financial costs of doing research with undergraduates and identify appropriate sources of funding. With funding in hand, you can then recruit undergraduates to the project and into your group. Most STEMM research labs take the first approach of coming up with the project idea (even perhaps with one or more students in mind) and then figuring out funding. But we think that the second approach—knowing the costs and having the funds in hand—is better for supporting and mentoring undergraduates in research. There are at least two reasons why this approach is preferable:

- **Diversity.** In securing the funding first (and not having a specific student in mind), you can have an open recruitment process (see Chap. 3). An open recruitment process can help you reduce unconscious or implicit bias while increasing the likelihood that you will bring diverse or unexpected students into your research group.
- **Equity.** By having the funding in hand, you have already determined the "wages" for your student researcher and how many students you can hire. You might think that having a budget in advance will unnecessarily constrain the research project. However, knowing your budget will reduce your risk of needing to hire more students than you budgeted for or paying them different amounts because you had to scrape together more money for the additional hire. Students talk to one another and discover any inequities; as morale collapses, so might the project.

Thus, we start our discussion of funding undergraduate research with an overview of the costs. But if you already know these costs, please skip ahead to Sect. 2.3 (defining projects) and Sect. 2.4 (finding the funding for them).

2.2 The Costs of Undergraduate Research

There are many different financial costs for undergraduate research, not all of which will be incurred for every student working in your research group. If you're just learning about these costs, a great place to start is with the categories and guidance used when creating a budget for a grant proposal: salaries/stipends and fringe benefits; subsistence (including housing and meals); travel; "other direct costs" such as research supplies, bench fees, server space or cloud storage, etc.; and indirect costs (*a.k.a.* "overhead").[1]

2.2.1 Stipend or Salary, Credit or Volunteer?

If you are supporting an undergraduate in a summer, intersession, or academic term research experience or internship in the lab or field, you should pay them a stipend or a salary.[2] Traditionally, stipends are meant to cover basic costs (i.e., room and board), but undergraduate researchers should be paid stipends that reflect a "living" (as opposed to "minimum") wage.[3] Underpaying students simply because they are interns or trainees who are not subject to minimum wage laws perpetuates inequities in STEMM. If you think that your mentees are just students who need the experience more than the pay, you are likely to discriminate against students from low-income backgrounds and others who need paying jobs to be able to attend and stay in school.

In contrast, undergraduates doing mentored research with you as part of a for-credit independent research project (e.g., a senior thesis) normally are not paid a stipend or salary for doing such work. These students receive college credit (and perhaps disciplinary honors) that they pay for through their tuition.

Some students may be willing to volunteer to be part of your research group. In general, we discourage taking on volunteer researchers. You are paid for doing research, and your students should be paid (or receive course credits), too. Your student-researchers also are more likely to take and maintain responsibility for their research when they're paid for it, and likewise, you can, and are more likely to, hold them accountable for their work and performance. Finally, and perhaps most importantly, students who can afford to volunteer to do research are a biased and usually less diverse subset of those who are already interested in, or could be brought into, STEMM fields (Fournier and Bond, 2015; Freund, 2017; Hance, 2017).

2.2.2 Subsistence Costs

Just like you have to pay your mortgage or rent and buy food, your undergraduate researchers do, too. During summer, intersession, or term-time, there are always students who need to find housing on the open (rental) market and buy and cook their own meals. If your student-researcher is working with you during the summer or intersession, they might be able to get dormitory accommodations and a meal plan from the institution. You should, however, check where or to whom those costs are billed and plan accordingly with the student. If your student-researcher is working with you during term-time, you might assume that their room and board are somehow taken care of by the college or university, but this may not be the case for many students (e.g., those who commute from home, live off-campus, or are so-called "non-traditional" students—those who are older than the average undergraduate, attending school after military or civilian service, or have different levels of family responsibilities).

You might also think that your student will be able to pay for their room and board out of their stipend, but we caution against thinking that stipends can or should be used to cover subsistence costs. For some students, this may be the case, but for many others, this is not a reasonable expectation; they may need their stipend to cover additional, equally important expenses.[4] If you do expect your students to pay for room or board from their stipend, be clear about this expectation and factor those expenses into their stipend. It is incumbent on you as a mentor to be aware of economic differences and needs among potential mentees. At the same time, you should not recruit students into your research group based on their ability to pay their own way.

2.2.3 Travel Costs

There are many different kinds of travel costs. Field researchers have to travel to and from field sites or field stations, and it is expected that these costs (gas, mileage) will be incurred by the research lab or project for everyone on the team. Lab-based researchers who spend summers or intersessions at off-site institutional or government laboratories and facilities (e.g., Brookhaven National Lab, Cold Spring Harbor, The European Center for Nuclear Research (CERN), NASA's Jet Propulsion Lab (JPL), Woods Hole) also expect travel costs to and from the lab to be covered. Similarly, travel expenses (to field locations or other labs) for undergraduate researchers should be covered by the project or research lab. When working with students from and at their home institutions during the summer or intersession, you also should recognize that your students may have gone home after the term ended and need to travel again to get back to the lab or to field sites.[5]

You shouldn't be expected to pay for your student's costs of commuting to and from the lab, but research-related travel should be covered, including initial and final travel to the lab or institution if they are not local. Additionally, don't ask your students to pay for these costs out of pocket and then get reimbursed. Besides the fact that most students can't afford to do that or would end up paying unacceptably high amounts of credit-card interest, working with your institution's travel office or using institutional accounts with online travel companies can be substantially less expensive.

Finally, consider budgeting for travel for your undergraduate student-researchers to attend regional or national meetings where they can present their research.[6] As we discuss in Chap. 8, attending meetings are an integral part of your work as a researcher and they can be a formative experience for your undergraduate student-researchers, too.

2.2.4 Other Direct Costs

Your students' research probably will use the same supplies and equipment that you use—and pay for—to do your research. These can include lab reagents, sequencing kits, field supplies, memory sticks, cloud-storage fees, and the like. More researchers means more disposable supplies, so budget accordingly.

2.2.5 Administrative and Other Hidden Costs

You are likely to have few additional administrative costs if you're mentoring only one or two undergraduate student-researchers (there are, however, many more for undergraduate research programs; see Sect. 2.5 in this chapter). But it is important to ask your department chair or your administrative team if there are any administrative costs and how those are paid for. You should also recognize and acknowledge that mentoring student-researchers takes real time for which you are unlikely to be financially compensated. Rather than resent this, consider the time and energy that your mentors spent with you at the beginning of your journey in STEMM. This is how you pay it forward.

2.3 Defining Mentored Research Projects: How Much Student Independence Can You Afford?

In Chap. 1 of this book, we asserted that as mentors, we see potential in our mentees to become self-aware, independent-minded scientists and scientifically literate members of their communities. In working with our students—whom we hope will become mentees—on mentored research projects, our primary goal should be to encourage the development of their self-awareness and independence as scientists. Although this is an admirable goal to strive for, we have to continually ask ourselves what does "independence as a scientist" mean for a student just learning to do research and how much independence can we actually afford, in financial or other terms.

2.3.1 From First-Time Researchers to Self-Aware and Independent Mentees

Undergraduates engage in research experiences along a continuum. For students who aspire to be scientists but have never been in a research lab or field site before, learning to use a pipettor, a balance, or band a bird for the first time can be dazzling. Students who already have a substantial technical skill set might reap similar rewards from being part of a team working on open-ended questions within a framework you as a mentor have already defined. But if you define the project and hire a student into it, their opportunity to exert some independence and "take ownership" of the project may be lessened. Finally, students who have already taken many classes, worked in a lab, and been on a research team may not feel satisfied in an undergraduate research experience unless they can ask their own questions, perhaps write a short proposal, and then do the proposed work. As their mentor, you should both model these skills and behaviors while guiding them through these different levels as they learn and define for themselves what success in each means.

2.3.2 Can You Afford Independent Students?

How much of your students' intellectual independence can you afford? If your research program is grant funded, you've already framed the project and defined its objectives and goals. You've got to make sure the experiments are run, the observations are made on time, and papers reporting the results are written so that you've got enough material to get the grant renewed in a few years' time. If you have hired an undergraduate researcher or two, it may be more expedient to pay them to wash glassware or count colonies *ad infinitum* so your graduate students or post-docs can do the intermediate work (or their own spin-off projects). If you're working primarily with undergraduate researchers, you may still be under pressure to get the grant-funded work done. In all of these cases, you may feel that dedicating scarce grant funds to pay undergraduates to think for themselves or develop their own projects may not be the wisest career move.

In short, the source of funding can directly influence the scope and direction of research. You are more likely to be a mentor to your student—co-creating and co-developing a project with them (see Chap. 1)—if the funding is not tied directly to a particular project.

2.4 Finding Funding for Mentored Student-Research Projects

You will likely find it to be as easy (or as difficult) to find support for undergraduates to work on projects you've framed and predefined as it is to find support for undergraduates to work independently. The sources are certainly different, but the success rates are similar. The real difference, as we outlined above in Sect. 2.3, is which direction you want to go with any particular project or proposal.

2.4.1 Finding Funding for Mentor-Defined Projects

As a PI, you likely have written grant proposals to support your research. You can always include support for undergraduates in these grants (as salaried employees, work-study students, or interns receiving stipends and other "participant support costs;" Sect. 2.2). This is a great way to support undergraduate research, can be very cost-effective,[7] and can be part of the "Broader Impacts" section of research grants when it is required. And if you've budgeted these costs up front into your proposal, you've already identified, considered, and understood the actual costs of working with undergraduate researchers. Finally, as we discussed earlier in this Chapter (Sect. 2.1), knowing these costs up front and having budgeted for them makes it easier to recruit a diverse set of students into your inclusive research group.

2.4.2 Finding Funding for Independent Mentees

Pursuing new ideas and engaging students in defining their research can be a rewarding and successful way to combine teaching and research. With gentle guidance or big-picture framing, independent student projects can act as pilot projects that jump-start larger or longer-term research programs for the student or for you (Vignette 2.1). Most institutions of higher education, from baccalaureate colleges (a.k.a. primarily undergraduate institutions, or "PUIs") to research-driven universities have internal funds dedicated for mentored independent student research or that provide seed money for faculty to initiate new projects. Undergraduates may be required to write some of the former type of proposals, whereas co-writing seed-money or pilot proposals with your students can be a great way to teach students about the proposal-writing process (see Sect. 6.3 in Chap. 6). Even if the proposals aren't funded or the pilot projects don't work out, there are still important learning gains to be found along the way—for both mentees and mentors.

Vignette 2.1   Leveraging undergraduate research to launch a research program

Contributed by Aaron M. Ellison

I began my academic career as an Assistant Professor at Mount Holyoke College, a four-year liberal arts college in western Massachusetts. As a postdoc and then for my first few years as a professor, I had been been studying marine invertebrate communities of red mangroves in Belize, often supporting undergraduate travel and research (Merkt and Ellison, 1998 is an example of a paper published by an undergraduate involved in this research). When tenure looked like a real possibility, I became interested in establishing long-term research in Massachusetts that could more easily involve undergraduates. To get new research going while teaching multiple courses a term, I engaged three undergraduates in mentored research.

Using my department's work-study funds, I hired three undergraduates to help identify potential study systems to address mangrove-inspired questions. For the first semester, each student was tasked with reviewing ecological literature on nearby dynamic habitats—tussocks, sponges, and pitcher plants—and writing a basic literature review. I provided the students with broad framing and a general direction. The next semester, they were invited to continue on as work-study students to develop an independent research project that could be conducted during the summer months. Meanwhile, I worked with the department chair and Dean's office to identify internal funding to support the implementation of their research proposals for the upcoming summer.

By the end of the summer we learned that each study system had different limitations. Tussocks and their plant associated were temporally asynchronous (LaDeau and Ellison, 1999).[8] The sponges in the campus streams were haunted by snapping turtles and frequently damaged by storms.[9] But pitcher plants were just right. Both the plant and its food web could be monitored regularly and non-destructively throughout the summer and studied in both the lab and the field (Ellison et al., 2003).[10] The data from my students' summer research convinced me, that long-term experimental studies using pitcher plants would be feasible. I continued to work on this system through 2020 (Ellison and Gotelli, 2021).[11]

There also are external grant opportunities that can support mentored student research. For example, in addition to supporting experiential research programs such as Research Experience for Undergraduate (REU) Sites, many directorates and divisions at the US National Science Foundation provide supplementary support to current grantees for 1–2 mentored undergraduates per year. These REU Supplements are intended to support students who are from groups traditionally underrepresented in the sciences, including minoritized and disabled students. Your local grants or sponsored research office will have and maintains current lists of public and private funding opportunities that could help you support undergraduates in your research group.

Even though you may only support a small number of students this way in any given year, over time the impact is magnified many times over. The often unappreciated importance of this magnifying effect was expressed eloquently by the poet Louise Glück in her 2020 Nobel address (Glück, 2021):

> Those of us who write books presumably wish to reach many. But some poets do not see reaching many in spatial terms, as in the filled auditorium. They see reaching many temporally, sequentially, many over time, into the future, but in some profound ways these readers always come singly, one by one.

So too, our students and mentees.

2.5 From Mentors to Program Directors: Funding a Whole Program

In Sect. 1.5 of Chap. 1, we discussed why you might be interested in developing a more formal undergraduate research program and providing mentored research opportunities for more students than can fit in one research group. A central part of any undergraduate research program is its budget, and the budget should reflect your—and your institution's or organization's—mission and vision for research and education, and values with respect to diversity, equity, inclusion, and belonging of students, mentors, and support staff.

2.5.1 It Really Does Take a Village

As the leader of an undergraduate research program, you set the tone: define its scientific scope, educational objectives, and recruitment and participation goals. But rarely will you make it happen entirely on your own. You likely will need help identifying individual mentors; recruiting students; managing and evaluating applications; ensuring students arrive on time; are paid, fed, and housed; get required trainings (e.g., lab or field safety, appropriate conduct, research ethics); have additional educational workshops or seminars; and are knit into a cohesive community of researchers who will clearly benefit from the undergraduate research experience (see Chap. 4 and Vignette 4.1). In our own experience, the total administrative and support costs per student range from 50–100% of the total compensation (stipend, housing, food, travel) paid to each student. That is, if you budget $100,000/year for the research done by students in your undergraduate research program, the actual cost is more likely $150,000–$200,000. Where do all those additional dollars go?

The most successful undergraduate research programs include not only research but also educational components and attention to building a community of scientists.[12] Budget for these, too. Workshops and seminars are rarely expensive, but it is reasonable to provide modest honoraria for workshop facilitators or seminar speakers. Knitting groups of students into cohesive and inclusive communities takes more work. Depending on the length of your program, consider hiring a seasonal or part-time assistant who is nearer the age of the students, has had undergraduate research experience themselves, and is interested in, and can focus on, networking and community organizing.

2.5.2 Supporting the Village

Support and administrative costs vary greatly among institutions and organizations, but it is incumbent on you as the program leader to know what they are and budget accordingly. Whether you "borrow" time from existing staff or hire new staff for the duration of the grant or the part of the year when the program runs (e.g., for summer or intersession programs), your program budget should include these costs either as direct charges to the grant, as part of the indirect costs (overhead), or through a cost-share agreement with your department, division, provost, or CFO.

Mentors in your undergraduate research programs need support too. You can certainly argue that supporting one or more undergraduate researchers for each participating mentor has monetary value, but every mentor may not agree with you. Some granting agencies may allow you to budget for mentor stipends, but others will not.[13] Alternative, creative and allowable solutions might include purchasing additional lab or field supplies, paying mentors' travel costs to attend meetings with undergraduate mentees, or covering publication costs of papers published with undergraduate co-authors.

2.5.3 Ensuring Equity

It is crucial that all students participating in your undergraduate research program receive equivalent levels of support. Stipends should be the same for every student in the program. As many undergraduates are expected to contribute a certain portion of their earnings to their tuition costs, it is best if they take home all of their stipends. Housing, food, and travel costs may vary among students, and ideally should be handled directly between the program and the vendors (e.g., housing office, dining commons, travel agency). If your program is supported by multiple awards or multiple funding sources or agencies, it is likely that levels of support/student will vary among your grants. Be prepared to make up the difference to ensure equity, but ensure that the methods you use are allowable by the funder.[14]

2.6   Take-Home Messages

✔ Money is equity.
✔ Find the balance between fostering independent undergraduate research and working with undergraduates to get your own research done.
✔ Get creative with finding and using resources to support collaborations with students.

References

Ellison, A. M., & Gotelli, N. J. (2021). *Scaling in ecology with a model system.* Monographs in population biology (Vol. 64). Princeton: Princeton University Press.

Ellison, A. M., Gotelli, N. J., Brewer, J. S., Cochran-Stafira, L., Kneitel, J., Miller, T. E., et al. (2003). The evolutionary ecology of carnivorous plants. *Advances in Ecological Research, 33,* 1–74.

Fournier, A. M. V., & Bond, A. L. (2015). Volunteer field technicians are bad for wildlife ecology. *Wildlife Society Bulletin, 29,* 819–821.

Freund, C. (2017). The hidden costs of fieldwork are making science less diverse. Retrieved December 01, 2021, from https://www.salon.com/2017/12/25/the-hidden-costs-of-fieldwork-are-making-science-less-diverse_partner/.

Glück, L. (2021). The poet and the reader. *The New York Review of Books, 58*(1), 29.

Hance, J. (2017). A rich person's profession? Young conservationists struggle to make it. Retrieved December 01, 2021, from https://news.mongabay.com/2017/08/a-rich-persons-profession-young-conservationists-struggle-to-make-it/.

LaDeau, S. L., & Ellison, A. M. (1999). Seed bank composition of a northeastern U.S. tussock swamp. *Wetlands, 19,* 255–261.

Merkt, R. E., & Ellison, A. M. (1998). Geographic and habitat-specific morphological variation of Littoraria (Littorinopsis) angulifera (lamarck, 1822). *Malacologia, 40,* 279–295.

Chapter 3
Recruiting and Selecting Students

Abstract This chapter discusses recruiting and hiring diverse undergraduate researchers, with an emphasis on "recruiting for potential." Even a relatively short undergraduate research experience can make an enormous difference to a student who has a lot of potential but still has not settled on a career path. Each student doesn't present or express their interests, potentials, or abilities in the same way. Thus, the importance of setting priorities, successful outreach strategies for reaching a broad and diverse applicant pool, and understanding and reducing implicit bias in the selection of participants is emphasized. Best practices for reviewing applications and conducting interviews fairly also are presented. These concepts and ideas apply both individual students and to entire cohorts of formal undergraduate research programs.

In this chapter, we discuss recruiting and hiring diverse undergraduate researchers, with an emphasis on "recruiting for potential." Even a relatively short undergraduate research experience can make an enormous difference to a student who has a lot of potential but still has not settled on a career path. Because each student doesn't present or express their interests, potentials, or abilities in the same way, we emphasize the importance of setting priorities, successful outreach strategies for reaching a broad and diverse applicant pool, and understanding and reducing implicit bias in the selection of participants. We also present some best practices for reviewing applications and conducting interviews fairly. We extend these concepts and ideas from individual mentors to program leaders and talk about how to apply them to both individual students and to entire cohorts.

This chapter has one Text Box and one Vignette. Box 3.1 provides pointers to resources on how to identify and minimize implicit bias in yourself and in your fellow mentors. Vignette 3.1 illustrates two different types of potential in prospective undergraduate researchers and successful outcomes of recruiting them.

© The Author(s), under exclusive license to Springer Nature Switzerland AG 2022   23
A. M. Ellison and M. V. Patel, *Success in Mentoring Your Student Researchers*,
SpringerBriefs in Education,
https://doi.org/10.1007/978-3-031-06645-0_3

3.1   Setting Priorities

You've committed to being a mentor for one or more undergraduate researchers (Chap. 1) and you've secured funding for them (Chap. 2). The next step is identifying and recruiting undergraduates who could join your research team. Before you start recruiting students, it's a good idea to clarify your expectations and identify your priorities for the undergraduates on your team. You may have sketched these out in the proposal(s) you wrote to support the students, and some expectations and priorities may be set by your funders, but there is plenty more to think about.

Most undergraduates will work in your research group for a single summer, semester, or academic year. What can you reasonably and realistically expect from these 8–28 weeks of mentored research? You may think first of collecting more data from existing, ongoing projects or running a small pilot for a new project. These are low-risk activities that can give undergraduates an introduction and exposure to research while providing needed labor to support yourself, your graduate students, or your postdocs. But as an undergraduate research mentor, you should have other priorities, too. You can spark and cultivate a student's interest in the field and launch them on a career path in STEMM. You may also want to develop mentorship skills and qualities in other members of your research group. Before continuing on, take a break from reading, make a list of your expectations for your students, and prioritize them.

My expectations as a mentor, ordered from highest to lowest priority
1.
2.
3.
4.

...

We emphasize that while you are working to achieve any of your expectations, you should be actively working to diversify and build an inclusive community of STEMM researchers. Diversity can be defined in many different ways and can include race, ethnicity, gender, economic background, geography, and culture, among many other important and intersecting aspects. (Mack et al., 2014; Martínez and Gayfield, 2019). Further, inclusion and belonging doesn't happen automatically in a diverse research team (Tatum, 2017; Shavit and Ellison, 2021). Weaving and integrating diversity and inclusion into your explicit priorities for research and education can also help you identify your own implicit biases (Box 3.1) and reduce their impact as you recruit and bring undergraduate researchers into your research team.

Box 3.1   Implicit bias: recognize it and reduce it

We all have implicit bias. What is it and how can we get rid of it?

Let's start with some definitions.[1]

- **Bias** consists of attitudes, behaviors, and actions that are prejudiced in favor of or against one person or group compared to another.
- **Implicit bias** is a form of bias that occurs automatically and unintentionally, but nevertheless affects judgments, decisions, and behaviors.

The bad news is that you can't be treated for, or cure yourself of, your implicit biases. But the good news is that there are a range of approaches that you can use to recognize and work to minimize your implicit biases.

The most widely used, well established, and best validated test for implicit bias is the Implicit Association Test (IAT). The IAT was developed by Project Implicit and is available online for anyone to take. Once you know what your implicit biases are, you should explore the range of "best practices" for minimizing them (Galinsky and Moskowitz, 2000; Blair et al., 2001; Stewart and Payne, 2008; Girod et al., 2016).

But no one solution will work in all cases. Assuming that every student application should be judged in the same way is not the best way to minimize your implicit biases.

In summary, we suggest an iterative three-step process:

1. Use the IAT to identify and recognize your implicit biases;
2. Work with a group of diverse people who have diverse perspectives, and who you trust and respect when they call you out on your biases;
3. Be open to critique, learn from others, and improve your evaluative practices.
   And repeat …

3.2   Recruitment

Recruiting undergraduate researchers can range from hand-picking the "rock-star" students from your classes to inviting formal applications from around the world. We strongly discourage hand-picking and strongly encourage using formal recruitment processes and applications tailored to individual or programmatic needs and priorities. Take off your own blinders and recruit for potential.

3.2.1   Why Not Hand-Pick a Rock Star?

Why not simply offer a research opportunity in your group to one of the top students in your advanced undergraduate course in a STEMM discipline? Such hand-picking is easy, quick, and gets you a student you know and who will likely fit into your research group. However, unless your only priority as a mentor is to accelerate the trajectory of the already successful advanced student, selecting the top student in your class will not help you achieve your other expectations and priorities.

You may also think that hand-picking students is an effective way to increase diversity in your research group. However, absent a formal recruitment and application process, hand-picked minoritized students may feel singled out and tokenized. This feeling can be exacerbated if only one (or a small number) of faculty or staff in a given department hand-pick minoritized students. Such students may hear from others in the department that they only got a research position because you actively hand-pick minoritized students. Perhaps more insidiously, your colleagues also may feel that if you are going out of your way to hand-pick minoritized students, then they don't need to examine their own biases and work together for greater diversity and inclusion in the department, division, or school.

Vignette 3.1   Successful outcomes from recruiting for potential

Contributed by Robert Bell & Frederi Viens

Which applicants are selected for REU programs? Our ideal participant is a student at an "early stage" of their study of mathematics, especially those not already intending a PhD in mathematics. The students we write about were participants in the 2019 Summer Undergraduate Research Institute in Experimental Mathematics REU program at Michigan State University.

Maggie applied as a rising senior from a large research university; she was interested in studying economics, not mathematics. Her project, subsistence farming in the Lake Chad Basin, had an economics angle. This topic required strong skills in mathematics to understand and develop machine-learning methods, and Maggie studied the algorithms and why they work, and she appreciated the motivation behind the questions. She led her team's development of computational Bayesian methodology, and her clarity about the subject rivaled and exceeded anything her mentor had heard on the topic, including from scholars with decades of experience. Maggie now works at the US Federal Reserve Bank in St. Louis.

Olu applied as a rising junior from a regional community college; he was interested in pursuing a Masters degree and going into teaching. At the start of the REU, Olu was independently working through an entire book on mathematical reasoning and proof writing, studying

---

these topics for the first time. His independence and work ethic immediately stood out. A few weeks into the program, he began self-study of
more abstract topics. He immersed himself in the research environment,
working in a small team, and giving weekly presentations. Olu thrived:
a good listener, supportive teammate, and courageous sharer of ideas.
After the REU ended, he transferred to a four-year university, graduated
with honors, and now is a first-year PhD student in mathematics.

It has been inspiring to play small parts in their journeys. For us, these
are the successes that recruiting for potential lead to in a mathematics
REU program.

3.2.2 The Value of a Formal Recruitment Process

A transparent and formal recruitment process is a better route to identifying students with high potential for success and to diversify the STEMM community. We emphasize the importance of *recruiting for potential* because even a relatively short undergraduate research experience can change what a student chooses to study or pursues as a career. Recruiting for potential also encourages you to explore differences in students' interests, potentials, or abilities. These differences may be more pronounced among individuals in various demographic and diversity groups that are not well represented in different STEMM fields. These differences also may not be apparent to you because of your own background, demographic characteristics, and implicit biases.

It will take you some additional time and effort to develop a formal process for recruiting and hiring undergraduate researchers. But in the long run, the gains are worth the time and effort. You will be able to re-use and more rapidly implement your recruitment process the next time you want to recruit student-researchers. And rather than reinforcing your own network of connections, by reaching out to a much larger pool of potential students you will expand your professional network, too.

3.2.3 The Elements of a Formal Recruitment Process

The two key elements of a formal recruitment process are knowing and targeting your **applicant pool**; and **consistency, equity, and transparency** in messaging, position requirements, and deadlines. Each of these should be informed by and focused on meeting your own expectations and priorities for involving undergraduates in your research.

Applicant Pool

Who do you want in your research group? There is no "right" answer to this question. Your answers should reflect your own values and aspirations, and your goals for building and sustaining a research group are likely to change from project to project and through time. Whoever is funding your research also may have implicit or explicit student recruitment goals. These may complement or conflict with your own values or goals; articulating and balancing such trade-offs is an essential component of defining your applicant pool.

In our experience, most would-be and seasoned mentors want "the best" students, but "the best" can be the most curious, the most undecided, those who already have all the skills needed, the ones with demonstrated potential, etc. Individuals in each of these groups will look (or not actively look) for research experiences in very different ways. Advanced or highly skilled students who are already committed to a STEMM career are likely to look for and see opportunities where you may be most likely to post them: on departmental or national job e-boards, list-servers, and social-media feeds used by professionals in the field. These students also will respond quickly to a prompt from a professor or department chair to apply to a lab or program run by one of their friends or colleagues.

At the other extreme are unsure or insecure students with great but undeveloped potential. How do you reach them? Standard methods of advertisements and postings are unlikely to be effective at convincing these students to apply to work with you. Rather than waiting for these students to "come to you;" it is much more effective for you to go to them. For example, if you are interested in first- or second-year students at your own institution, drop in on the introductory courses (in the relevant STEMM discipline) or their associated labs and talk to the students about research opportunities. Broaden your horizons with a trip to a local community college, HBCU, or tribal college. Ask a professor or department chair there if you can give a talk about research opportunities in your lab or your department. You will be surprised how many will respond enthusiastically and affirmatively.[2]

Finally, take advantage of other lab alumnae/i or students who are currently working in your research group. They have their own networks that can help you build a diverse applicant pool.

Consistency, Equity, and Transparency

Be crystal clear in your messaging because how you communicate your priorities plays a vital role in recruiting applicants. To recruit a diverse applicant pool, define and emphasize diversity in your recruitment materials. If you are recruiting for potential, say so. If you want particular applicants to have specific skills, list them.

Most applications ask for specific, required materials. These include things like a cover letter, résumé, transcripts, personal essay(s), and letters of recommendation.[3] But don't require application materials that you have no intention of using or that perpetuates inequities. For example, if you want to know what STEMM courses a student has taken, it is enough to ask them to provide a list of their courses. A transcript might be needed only if one of your selection criteria is a particular set of grades or grade-point average.[4]

Finally, if you have an application deadline—for both applicants and recommenders (if you're asking for letters of recommendation)—stick to it. A key element of equity is consistency in the process: the same rules should apply to every applicant. Enforcing deadlines is made easier if you use an online application system that you can set to close at a specific time and date.

3.3 Reviewing Applications

You should apply the same consistent, equitable, and transparent methods to reviewing applications. Having and applying a clear set of standard criteria and characteristics for the students you want working in your research group will help you to minimize your own biases as you review a set of applications. Such standard "rubrics" are widely used in hiring decisions in the government and private sectors, but have a shorter history of use in academic institutions. Cross-check your rubric with your recruitment materials and methods. Your priorities for screening applications should have been reflected in your recruitment and messaging. For example, if you don't care about an applicant's grades or grade-point average, don't require transcripts in

Table 3.1 Different rubrics that help you recruit for students with past success in STEMM or great potential for future success in STEMM careers. What kind of students are you likely to accept into your research group if you use each of these two rubrics?

Topic	(Re)Framing the Question	
	Classic	Recruiting for potential
Coursework	What relevant STEMM courses has the student taken?	What courses did the student take based on the opportunities they had?
GPA (if it matters)	What is the student's GPA?	How has the student's GPA changed?
Experience	Does the student have experience in the relevant STEMM field?	What challenges may have affected the student's experience in STEMM?
Letters of recommendation	Who wrote the letters of recommendation and how strong are they?	What do the letters of recommendation say about the student's potential and ability to meet and overcome challenges?
Personal statement	How "well-written" is the personal statement or essay?	What does the personal statement or essay actually say about the student?

the application package. If you do ask for transcripts, though, it signals to potential applicants that you care about grades and you will be using them as one of your review criteria. Once you have developed a rubric for screening applications, you should use it consistently; don't change paths or make exceptions.

Application reviews and rubrics can maintain the *status quo* or effect change. In Table 3.1, we contrast a "classic" set of questions you might use to evaluate a student's application with re-framed questions you could use to increase diversity and inclusion and recruit for potential instead of past success.

3.4 Interviewing Applicants

You've done a great job recruiting a deep applicant pool, carefully reviewed the applications, and created a short list using objective criteria designed to minimize implicit bias. Now, you're ready to interview the students. Like reviewing applications, interviewing individual candidates presents its own challenges and threats of implicit bias. In-person interviews present sensory cues that can trigger both positive and negative responses. Ensuring consistency across interviews in their tone, platform and setting, and questions can significantly reduce the impacts of implicit biases.

3.4.1 Set the Tone

You want the students to be serious about their research and to take this opportunity seriously. So, the interview should be like a "job interview," not an informal or casual conversation. There's really no downside to a formal interview. You're interested in learning about their interest in the research opportunity, whether they can seize the opportunity, and if they are likely to do a good job with it. Decide in advance how long the interview will be, let the students know, and stick to time.

3.4.2 Platform and Setting

If you're interviewing the students in person, conduct the interview in your office or a nearby conference room, not over a meal or in a bar or café. If you interview one student in person, interview them all in person, even if that means bringing in one or more candidates to your campus for an on-site interview (and paying the costs of their visit). If that seems unrealistic, interview all the students, even the local ones, by phone or video-conference.

If you're interviewing students remotely, use the same platform for all the interviews and be aware of differences in time zones.[5] You may want to use videoconferencing, but if even one of your short-listed candidates doesn't have good Internet access or requests a phone interview—and you agree to their request—then interview all the students using the phone.[6] Remember that you may have super-fast broadband at your college or university, but your applicants may not, or may not have access to a private video terminal. Sensitivity to differences in infrastructure signals to your applicants that you are aware of the pitfalls of the digital divide.

Interviews present other opportunities for you to think about your own biases. For example, if you have biases about physical appearances or attire, don't use video. Just as you wouldn't want to be disturbed during an in-person interview, don't do remote interviews in a distracting environment. For each interview, find a quiet location, select a mutually convenient time of day when demands before and after the interview are low, dress professionally, and focus on the interview (turn off your email and other open programs on your computer's desktop). The importance and seriousness of the research opportunity should be reflected in your demeanor and the interview.

3.4.3 Consistency in Questions

You should ask all the students the same set of questions so that when you're comparing the applicants later you have comparable information. We find it helpful to write down your interview as a dialogue or script, so that you stay on task and are consistent from person to person. You should be aware of local, state, and federal regulations regarding questions you can *and cannot* ask during an interview.[7] For many reasons, it's best not to record the interviews. Rather, you should take notes and write down key answers.

3.4.4 Wrapping up the Interview

There are a few important items that are crucial to completing an interview:

- Reprise and restate the position requirements, skills, logistics, expectations, and compensation. Be sure that the interviewee understands these requirements and expectations and is agreeable to them.
- Detail the timeline. Let each interviewee know when will you make a decision about which student(s) to take on, how will you communicate the decision (e.g., email, phone, text), and how long thereafter will you expect a response—positive or negative—from the student. Make that time long enough so that you can make an offer, have it accepted, and send polite rejection letters all by the expected date.
- Allow time for the interviewee to ask questions they might have about the opportunity.

3.5 Selecting and Hiring Students

If you have set up rubrics for reviewing applications and interviewing students, it should be straightforward to identify which among your candidates is the most qualified for the position. The successful candidate(s) should be excited about the research opportunity and be able to do the work expected in the time available. Concern about whether or not a student will "fit in" to the research group is an indicator of implicit bias. As a 21st-century mentor, you should be interested in building a diverse and inclusive research group, not perpetuating an insular one.

Follow up with the student you want to hire as quickly as possible. Since you concluded the interview by laying out a timeline, you should ensure that the time from offer to acceptance (or rejection) fits comfortably within it. The offer of a position should be in writing, and include all the position requirements, skills, logistics, expectations, and compensation that you included in the position description and that you discussed during the interview. Should disagreements arise later ("you never told me I would need to be pipetting for 24 h straight"), the offer letter can serve as a useful reminder.

Your college, university, company, or institution undoubtedly has rules and procedures for hiring a student, getting them ID cards, onto payroll, etc., and there may be further discussions about any necessary travel, housing, or food. If you are fortunate enough to have administrative assistance to help you with these logistics, take full advantage of it (and budget for it). If not, stay up-to-date on the rules and procedures and ensure all are handled in a timely fashion.

3.6 From Mentors to Program Directors: Recruiting New Mentors

In building and sustaining an experiential undergraduate research program, your goal should be to recruit and engage a cadre of diverse and inclusive mentors who are committed to research and education. All of the criteria that you would use to recruit students into your research group work equally well for recruiting mentors into a new or established programs. An individual's potential, including their enthusiasm to work with undergraduates and their willingness to learn the "soft" skills that distinguish a mentor from an advisor or supervisor, is, in our experience, much more important than the number of papers they've published or their status in the field.[8] Recruiting mentors for their potential establishes yourself as an inclusive leader (or mentor-of-mentors) and empowers them to learn new skills and pass on their own enthusiasm and empowerment to their mentees.

3.6.1   Research Mentors

You are most likely to look first to recruit established researchers who have had the opportunity to advise and mentor students and younger colleagues. However, you are likely to discover that few of them have had formal education in the practice of mentorship. This need not be a problem as long as they buy in to the overall programmatic goals and are willing to learn and use new mentoring skills.

Of course, research mentors need not all be established, senior scientists. Early-career researchers, especially graduate students and postdocs, bring new ideas and energy to undergraduate research programs. Although most graduate students and postdocs can probably identify someone who was an important mentor to them, they probably haven't thought that they could be a mentor themselves. You can point out that being a research mentor can help them improve their own research, create opportunities to develop new skills, and may even give them an extra edge when they're looking for a permanent position in any sector. And it's never too early to confront one's own implicit biases and contribute to diverse and inclusive STEMM communities.

3.6.2   Trainers and Facilitators

It is similarly important to have a diverse and inclusive group of professionals working with your program who provide necessary training courses and facilitate educational workshops for the students (see Chap. 6 of this book). Some of your research mentors may also be able to fill these roles (e.g., for training in research with animals or people, or workshops on formal presentations), but it is likely that you will need to bring in others who have specific skills and backgrounds. Talking with appropriate offices and groups (e.g., Environmental Health and Safety, Sponsored Research Office, Writing Center) on your campus or organization can point you toward people who can offer valuable training courses and workshops.

3.6.3   Support Staff to Support Diverse Students

Finally, there is very often a strong disconnection between the diversity, cultures, and backgrounds of more senior researchers and undergraduates. Despite your best efforts to diversify your pool of mentors, in most programs—especially those at large research institutions—research mentors tend to be much more homogeneous than the students. This is no excuse to not continue to recruit more diverse and inclusive mentors, but you should also ensure that there is adequate staff in your program and throughout your organization who can support your diverse group of undergraduate researchers.

3.6.4 Ensuring Equity

Most importantly, representation and responsibility for creating an inclusive, welcoming research and education environment should be spread as equally as possible throughout your program. We don't encourage program directors to bring a separate diversity and inclusion "expert" onto the team. Rather, everyone on the team should share your commitment to a diverse and inclusive program and bring their own lived experiences to the team of mentors.

3.7 Take-Home Messages

✔ Recruit for potential.
✔ Be aware of your biases and learn ways to minimize them.
✔ Formal is better than informal, especially when it comes to expectations, rubrics, application guidelines, and interviews.

References

Blair, I. V., Ma, J. E., & Lenton, A. P. (2001). Imagining stereotypes away: The moderation of implicit stereotypes through mental imagery. *Journal of Personality and Social Psychology, 81*(5), 828–841.

Galinsky, A. D., & Moskowitz, G. B. (2000). Perspective-taking: Decreasing stereotype expression, stereotype accessibility, and in-group favoritism. *Journal of Personality and Social Psychology, 78*(4), 708–724.

Girod, S., Fassiotto, M., Grewal, D., Ku, M. C., Sriram, N., Nosek, B. A., & Valantine, H. (2016). Reducing implicit gender leadership bias in academic medicine with an educational intervention. *Academic Medicine, 91*(8), 1143–1150.

Mack, K., Taylor, O., Cantor, N., & McDermott, P. (2014). If not now, when? The promise of STEM intersectionality in the twenty-first century. *Peer Review, 16*(2), 29–31. https://www.aacu.org/publications-research/periodicals/if-not-now-when-promise-stem-intersectionality-twenty-first.

Martínez, A., & Gayfield, A. (2019). The intersectionality of sex, race, and hispanic origin in the stem workforce. US Census Bureau Social, Economic, and Housing Statistics Division Working Paper 2018-27:1–24.

Shavit, A., & Ellison, A. M. (2021). Diversity is conflated with heterogeneity. *The Journal of Philosophy, 118*, 525–548.

Stewart, B. D., & Payne, B. K. (2008). Bringing automatic stereotyping under control: Implementation intentions as efficient means of thought control. *Personality and Social Psychology Bulletin, 34*(10), 1332–1345.

Tatum, B. (2017). *Why are all the black kids sitting together in the cafeteria? And other conversations about race* (Revised). New York: Basic Books.

Part II
Mentoring Student Researchers

The three chapters in this central section of this book consider the undergraduate research experience in STEMM as a "three-legged stool" whose legs are research, education, and becoming part of a professional community. For many undergraduate students, the process of doing research, learning new skills, and being part of a team or community of like-minded peers and role models can be a transformative experience. Becoming part of a community and learning to do research are not things that can be learned by reading books or being in a classroom. Technical and professional skills can be taught in classroom settings but are learned more effectively in hands-on, participatory workshops. What students learn from doing research, the technical and professional skills they acquire, and the community of scientists they work with, all contribute to a scientific literacy that they will carry with them on any career path.

Chapter 4
Building a Research Community

Abstract Being part of a professional community of collegial researchers is critical to long term success in any STEMM field. This first of three chapters on mentoring students through their undergraduate research experience focuses on how to build, organize, and manage an inclusive community of researchers. The emphasis in this chapter is on research communities that include one or more undergraduates or other new researchers in a single research group or in an undergraduate research program, but the principles described are applicable to any research group. The value of connecting one's mentees to networks of STEMM practitioners and professionals is of key importance. These networks extend well beyond the groups of graduate students, post-docs, and faculty studying and working in colleges and universities.

Being part of a professional community of collegial researchers is critical to long term success in any STEMM field (Wuchty et al., 2007). Thus, we begin our discussion of mentoring students through a research experience with a discussion of how to build, organize, and manage an inclusive community of researchers. We emphasize research communities that include one or more undergraduates or other new researchers in a single research group or in an undergraduate research program, but the principles we describe are applicable to any research group. We stress the value of connecting your mentees to networks of STEMM practitioners and professionals, of which there are many more than groups of graduate students, post-docs, and faculty studying and working in colleges and universities.

There is one Text Box and one Vignette in this chapter. Box 4.1 summarizes the key elements of successful research teams. Vignette 4.1 discusses best practices for building and maintaining diverse and inclusive research teams (for other examples, see Uriarte et al., 2007; Cheruvelil et al., 2014; van der Wal et al. 2021).

4.1 STEMM Communities

As a scientist who does research, who is in your professional community? And is there only one, or are you part of multiple, overlapping or non-overlapping communities (Guimerà et al., 2005)? For example, your primary professional community could be the students, research assistants, graduate students, or post-docs who make up your research group or "lab." It could be a small group of other researchers with whom you routinely co-author papers or research proposals, or a larger, more varied and less consistent group of other researchers with whom you collaborate. Your community could be centered on a scientific society for which you may be part of a section, chapter, or interest group; a journal editor; or a committee member or officer who sets directions for the field. It could be some or all of your colleagues in an industrial laboratory with a research arm, in a nonprofit organization using science to improve social well-being, or in a state or federal agency that make policies grounded in scientific evidence. STEMM communities can be local, regional, national, or international, and many of their members may interact only virtually and never meet in person.

Why is being part of a professional community so important for new researchers? A professional community with a common purpose instills a sense of belonging in its members, applauds their successes, and encourages them when they are less successful. When they feel a part of a professional STEMM community, students who haven't previously thought of themselves as being able to "do science" may learn that they are excited by the prospect of becoming a scientist. Students who have considered pursuing a STEMM career often solidify that goal after doing research for the first time. And of course, some students who thought that they would become scientists decide that they'd rather focus their efforts in other directions. Although some consider these students to have "left the field" or been "lost" from the STEMM "pipeline," we do not.

The pipeline metaphor implies a straight path through the educational system to a career in STEMM; this metaphor is pervasive in both the popular press and the technical literature. Further, Garbee (2017) illustrated clearly that the pipeline metaphor focuses on only one pathway to a particular kind of STEMM career (i.e., as a college or university professor) and emphasizes the negative: "leaks" or losses from the pipeline as individuals leave school, get different jobs, or change career directions. The pipeline metaphor is also based on a set of assumptions about future job growth in the STEMM sector driving economic growth (Institute of Medicine, 2007). Although the 2007 IOM report justifiably called attention to the lack of diversity in STEMM professionals in elementary, secondary, and post-secondary education, its focus on a single career path and losses from the pipeline was restrictive and the metaphor continues to be counterproductive (e.g., Garbee, 2017). More recent assessments and analyses have focused more constructively on a wider range of career paths and opportunities for individuals to move in and out of STEMM pathways and communities at different times in their lives (e.g., National Science Board, 2015).

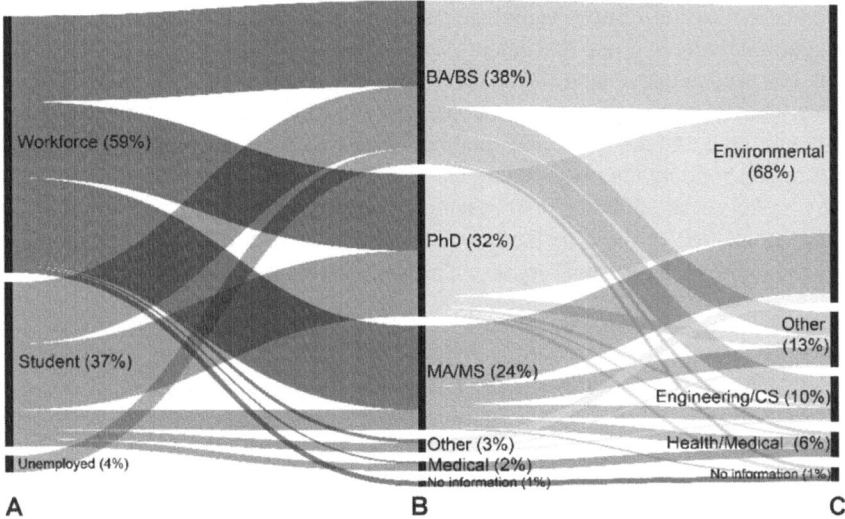

Fig. 4.1 An alluvial diagram showing the braided and varied careers of alumnae/i (1998–2018) from the Harvard Forest Summer Research Program. Of the 240 respondents (out of 512 alumnae/i) to our 2019–2020 survey on their education and career status in the years following their undergraduate research experience at Harvard Forest, most respondents are currently employed [**A**]; 58% earned or are currently pursuing a post-graduate degree (MSc or PhD) [**B**]; and most are working or studying in an area related to ecology or the environment (the focus of the Harvard Forest Summer Research Program) [**C**]. Data from McDevitt et al. (2020) and Ellison (2021); plot by Audrey Barker Plotkin

A better visualization than a leaky pipeline is a braided, alluvial stream (Fig. 4.1).[1] This figure illustrates that there are different career sectors, emphasizes diversity and variation in STEMM paths to success, and de-emphasizes a rigid temporal sequence of study and work. For example, the "workforce" category in Fig. 4.1 could be subdivided into a variety of specific jobs; the "student" category could be subdivided by demographic groups; and the "unemployed" category could reflect different reasons for being neither in the workforce nor in school. Different individuals in each sector have different levels of schooling or types of degrees, and individuals working or studying in different STEMM areas vary in their career paths. Finally, rather than focusing on "leaks" and "losses" from STEMM communities and professions, Fig. 4.1 simply identifies absence of information.

4.2 Building a Collaborative Research Team

For new researchers, and especially for students, their immediate research group is their most important professional community. As a group leader and research mentor, your responsibilities include not only defining the direction and intellectual content of the research effort but also providing the necessary support for it. Such support will be monetary, administrative, technical, and emotional. Your goal should be to build and guide a research group that can provide all facets of support. To accomplish this goal, you need to understand and be explicit to your group members what the project needs, and then to find and support people who can work together to make it happen. Although many of the ideas and methods discussed here are based on extensive research on building interdisciplinary teams of established, professional researchers (e.g., Stokols et al., 2008a, 2008b; Hampton and Parker, 2011; Pennington, 2011; Dahlin et al., 2019), these methods are as relevant to your own research group as they are to "working groups" organized around specific questions or syntheses.

4.2.1 Roles in a Professional Community are Not Static

There are different ways that individuals participate in research teams and how someone participates can change over time. Someone who started as a research assistant may become an intellectual leader. Someone who started in administrative support may evolve into an editor or co-author. Someone who started as an intellectual leader may realize that they need to return to the classroom to learn new skills. And someone who was once focused on attaining an advanced degree in a technical field may find that they're more adept at bringing people together and guiding the otherwise invisible social interactions that hold the group together. Acknowledging the important professional and social roles teams serve goes a long way to make them successful (Box 4.1).

Box 4.1 The twelve most important elements of successful research teams

Ten elements of successful research teams have been repeatedly identified (Bennett et al., 2010):

1. Trust
2. Vision
3. Self-awareness
4. Leadership
5. Mentoring
6. Allowing for evolution and dynamics
7. Open and honest communication
8. Sharing recognition and success
9. Taking advantage of conflict and disagreement to expand thinking and stimulate research
10. Navigating and leveraging broader networks

> To this list we would add two more:
>
> 11. Equity in support and expectations
> 12. Teaching and using practical skills
>
> Trust always comes first on any list. Your mentees will be part of a research team and of much broader communities of researchers. Everyone needs to trust themselves and others on the team, and trust the value of, and contribute fully to, group efforts.

4.2.2 Bringing New People into Your Research Team

Research teams do not spring fully-formed from a random collection of people. Rather, they need to be carefully assembled and cultivated (Vignette 4.1). For example, Guimerà et al. (2005) used data from teams of various sizes that had achieved varying degrees of success in the arts and sciences. They identified a process of self-assembly of teams that occurred across fields as a function of team size, the proportion of new individuals, and how likely members of the team were likely to repeat previous collaborations. Like Wuchty et al. (2007), Guimerà et al. (2005) found a steady increase in team size and performance over time, with more rapid increases in the sciences than in the arts or social sciences. They found that successful teams have a higher fraction of repeat collaborators ("incumbents") who bring continuity of process and knowledge, and that more diverse teams outperform less diverse ones (see also Institute of Medicine, 2007). Their results also suggested that there could be an optimal size for teams, but that this size is likely to vary among disciplines.

> **Vignette 4.1 Building diverse teams**
>
> *Contributed by Emilie Stander*
>
> I regularly hire students pursuing their associate's degrees as paid interns to assist in water quality and sustainability research and stewardship projects that I run on our college campus and in the local community. There are many students with applicable skills, such as critical thinking, written and oral communication, academic potential, and soft skills. Yet these same students often suffer from anxiety and other mental health challenges that cause them to frequently miss deadlines for class assignments, have difficulty completing projects, or suffer from a paralyzing lack of confidence. Many generations of students have quietly struggled with these challenges; some may have opted out of student research opportunities or did not feel included by mentors who sought out the most stellar, high functioning students. In colleges and universities, educators and mentors recognize the importance of sup-

porting students' mental health so that all students can make valuable contributions to scientific research.

By being an inclusive community and providing layers of support, we can benefit from what students have to offer. I work with students to find ways to play to their strengths, while providing guidance, mentorship, and compassion. I pair them with peers who can support them and hold them accountable in less formal ways. Through this team approach, some of these students have been able to shine in their internship roles. One student won an award from a local environmental non-profit for obtaining funding and organizing a riparian forest restoration project on our campus. Another developed outreach materials of such high quality that our college president and other prominent members of the community have taken notice. It doesn't always work out, but I have found it worth the effort to give these students a chance.

There will always be times when some people leave the team and others join, and these changes can be difficult and stressful. These changes and their attendant challenges are most apparent when welcoming new students into your team who have never done research before or are ambivalent about pursuing STEMM careers. For most of these students, participation in research teams can be intimidating or fail outright for many reasons, including:

- Insufficient preparation;
- A lack of clarity or misunderstanding about team structure;
- Unsure about "appropriate" personal conduct;
- Inexperience or shyness in speaking or sharing ideas to a group;
- Internalized cultural attitudes about speaking, identifying difficulties ("complaining"), confrontation, or talking about feelings;
- Other members dominating meetings or other avenues of communication, leading to a lack of equitable space or "airtime;"
- Too great a focus on efficiency;
- Lack of informal, lighthearted, or other social elements among team members.

You need to be aware of non-verbal cues that suggest that new (and more seasoned) members of your team are feeling unsure of themselves or intimidated. You need to actively invite, listen to, and follow through on suggestions from them for improvement. It is critical that you strike a balance between encouraging students towards participation and discouraging damaging outcomes.

4.2.3 The Importance of Formal and Informal Communication

Many studies have shown that teams with open and honest lines of communication are particularly effective. In formal team meetings, ensuring that everyone has the opportunity to share their ideas and opinions and that others hear and listen to them and treat them with respect demonstrates your own commitment to equity across your team. Dahlin et al. (2019) provides a set of tools for facilitating collaborative team meetings that help to ensure that everyone's voice is heard. These tools are as important and valuable in virtual meetings as they are in face-to-face ones.

But informal communication in unstructured breaks or other social interactions can be even more valuable. For example, Pentland (2010) presented results from a study in which he and his colleagues deployed wearable electronic sensors to track interpersonal interactions among co-workers in the private sector. They found that productivity increased up to 20% when coffee breaks for all staff were scheduled simultaneously.

Putting these ideas together, there are five key contributors to communication within successful teams (Pentland, 2010):

- Everyone on the team speaks and listens in roughly equal measure, keeping contributions short and sweet.
- Members face one another, and their conversations and gestures are energetic.
- Members connect directly with one another, not just with the team leader.
- Members carry on back-channel or side conversations within the team.
- Members periodically break, go exploring outside the team, and bring information back.

4.3   From Mentors to Program Directors: Developing Communities of Researchers

It's critically important that you develop and articulate your own vision, mission, and values for creating and building the various research, education, management, and support communities associated with your undergraduate research program. As the program director, you set the tone and working process for each of them. How you support your team of research mentors is likely to be reflected in how each of them supports their individual research teams. If you work effectively with your team of supporting personnel (e.g., individuals working in administration, logistics, or facilities), many of the day-to-day tasks that keep a program running smoothly should happen out of sight but not out of mind of the research groups. And, you will also be part of a network of research program directors working together to advance community-wide goals for diversity and inclusion in STEMM.

In each of these teams, you should always be using the twelve keys for successful research teams (Box 4.1). First and foremost, trust your colleagues. Respect the different mentoring styles of the research mentors in your program and don't micromanage their research or lab operations. Understand that your support staff help others besides you. Provide clear schedules of tasks that must be done with plenty of advance notice but also help them understand—and understand for yourself—that not every contingency can be predicted in advance. Keep in regular touch (at least weekly, but daily is better) with all of your different teams about individual successes and challenges. As a community leader, you should be receptive to suggestions for improvement or changes in direction and always be ready to pick up any slack in the system.

4.4   Take-Home Messages

✔ Diverse and inclusive STEMM communities come in many forms.
✔ Trust is the foundation for successful research groups.
✔ Informal interactions are as important as formal interactions for ensuring success in collaborative research groups.

References

Bennett, L. M., Gadlin, H., & Marchand, C. (2010). *Collaboration and team science: A field guide.* Bethesda: National Institutes of Health.

Cheruvelil, K. S., Soranno, P. A., Weathers, K. C., Hanson, P. C., Goring, S. J., Filstrup, C. T., & Read, E. K. (2014). Creating and maintaining high-performing collaborative research teams: The importance of diversity and interpersonal skills. *Frontiers in Ecology and the Environment, 12*(1), 31 38.

Dahlin, K. M., Zarnetske, P. L., & Record, S. (2019). Hear every voice - working groups that work. *Frontiers in Ecology and the Environment, 17*(9), 493–494.

Ellison, A. (2021). Harvard forest summer research program in ecology student surveys 2016-2021 ver 2. Environmental Data Initiative. Retrieved September 16, 2021.

Garbee, E. (2017). The problem with the "pipeline". Retrieved September 10, 2021, from https://slate.com/technology/2017/10/the-problem-with-the-pipeline-metaphor-in-stem-education.html.

Guimerà, R., Uzzi, B., Spiro, J., & Nunes Amaral, L. A. (2005). Team assembly mechanisms determine collaboration network structure and team performance. *Science, 308,* 697–702.

Hampton, S. E., & Parker, J. N. (2011). Collaboration and productivity in scientific synthesis. *BioScience, 61*(11), 900–910.

Institute of Medicine. (2007). *Rising above the gathering storm: Energizing and employing america for a brighter economic future.* Washington, DC: The National Academies Press.

McDevitt, A. L., Patel, M. V., & Ellison, A. M. (2020). Lessons and recommendations from three decades as an nsf reu site: A call for systems-based assessment. *Ecology & Evolution, 10,* 2710–2738.

National Science Board. (2015). *Revisiting the STEM Workforce, a companion to science and engineering indicators 2014.* Arlington: National Science Foundation.

Pennington, D. D. (2011). Collaborative, cross-disciplinary learning and co-emergent innovation in eScience teams. *Earth Science Informatics, 4,* 55–68.

Pentland, A. (2010). The new science of building great teams. *Harvard Business Review, 90,* 60–70.

Stokols, D., Hall, K. L., Taylor, B. K., & Moser, R. P. (2008a). The science of team science: Overview of the field and introduction to the supplement. *American Journal of Preventative Medicine, 35*(2S), S77–S89.

Stokols, D., Misra, S., Moser, R. P., Hall, K. L., & Taylor, B. K. (2008b). The ecology of team science: Understanding contextual influences on transdisciplinary collaboration. *American Journal of Preventative Medicine, 35*(2S), S96–S115.

Uriarte, M., Ewing, H. A., Eviner, V. T., & Weathers, K. C. (2007). Constructing a broader and more inclusive value system in science. *BioScience, 57*(1), 71–78.

van der Wal, J. E. M., Thorogood, R., & Horrocks, N. P. C. (2021). Collaboration enhances career progression in academic science, especially for female researchers. *Proceedings of the Royal Society B, 288,* 20210219.

Wuchty, S., Jones, B. F., & Uzzi, B. (2007). The increasing dominance of teams in production of knowledge. *Science, 316,* 2036–1039.

Chapter 5
Doing Research with Undergraduates

Abstract Doing research with undergraduates can be rewarding on many levels: from something as simple as introducing a student to the joys of asking a question to something as tangible as a successful co-authored paper. Yet, senior researchers often say that teaching and mentoring undergraduates in research prevent them from "getting their own work done." This chapter discusses how research mentors can accomplish their research goals while integrating undergraduate students into research program and mentoring them as researchers. Three points are emphasized: (1) the importance of setting expectations (for your mentees *and* yourself); (2) empowering mentees through feedback, encouragement, and allowing them to "take ownership" of a part of a larger research endeavor while preserving its conceptual integrity; and (3) working with mentees to help them identify opportunities for further intellectual and career growth in STEMM.

Doing research with undergraduates can be rewarding on many levels: from something as simple as introducing a student to the joys of asking a question to something as tangible as a successful co-authored paper. Yet, senior researchers often say that teaching and mentoring undergraduates in research prevent them from "getting their own work done." This statement is false on two counts. First, teaching and mentoring, like supervision of technicians, advising students, and other aspects of managing a research group, are defined and expected parts of our jobs. Second, and most importantly, teaching and mentoring strongly contribute to all aspects of research while expanding its scope and reach.

In this chapter we focus on how you can accomplish your research goals while integrating undergraduate students into your research program and mentoring them as researchers. We discuss setting expectations (for your mentees *and* yourself); empowering your mentees through feedback, encouragement, and allowing them to "take ownership" of a part of a larger research endeavor while preserving its conceptual integrity; and working with mentees to help them identify opportunities for further intellectual and career growth in STEMM. This chapter has one Vignette, which illustrates the challenges of working with students who are not communicating or meeting expectations (Vignette 5.1).

A. M. Ellison and M. V. Patel, *Success in Mentoring Your Student Researchers*,
SpringerBriefs in Education,
https://doi.org/10.1007/978-3-031-06645-0_5

5.1   Setting Expectations

You've gotten to the point of defining and funding a research project that can include undergraduates (Chaps. 1 and 2) and recruited one or more student researchers (Chap. 3). Now, they are at your door and ready to get started. The students are curious, open to new ideas and directions, and ready to work. Keep in mind that undergraduate researchers are neither technicians or research assistants doing clearly defined jobs, nor are they graduate students or post-docs with years of experience in the field. But just as you have clear expectations as a supervisor or manager for your employees and clear expectations as an advisor for your graduate students and postdocs, you should have clear expectations for your undergraduate mentees that lay out the priorities that you and they jointly agree on.

There's no recipe for setting and communicating expectations for undergraduate research that will apply in every situation or to every mentee. But there are general principles that can help you to develop a set of expectations and a system for communicating and following through with them that works for everyone concerned.

A good set of expectations combines what you want from the mentees and what they want from you,[1] and what you can give to them and what they can give back to you. These expectations should always be placed in the context of the goals, objectives, and requirements of your overall research program. They should also reflect the normally short time available for the undergraduate research experience. There also may be additional expectations or constraints defined by who is funding the research. You should anticipate that not everything will go according to plan, so you should have the ability to redefine or change expectations when you need to. Perhaps most importantly, expectations should be communicated clearly and explicitly.

5.1.1   The Research Comes First, But it is First Among Equals

An undergraduate research experience is, first and foremost, about learning how to do, and doing, *research*. Undergraduates participating in research experiences have different ideas of what research actually is.[2] From the start, you should ensure that your undergraduate mentees understand what research in your group entails, how it differs from what they may have seen in previous coursework, and how open-ended it is likely to be. You probably have your own short- and long-term goals for your research program. You should make your mentees aware of these goals and help them understand how their own research activities will contribute to these goals.

Research does not happen in a vacuum. There are technical and professional skills that all scientists need to learn and refine to do their research effectively and ethically. Most undergraduate research experiences therefore include educational components (e.g., training sessions, workshops, seminars, post-graduate education and career planning, etc.; see Chap. 6). You should ensure that your mentees have access to these educational components, be clear about how essential they are to doing research successfully, and whether attendance and participation in them is mandatory or optional.

Similarly, successful scientific research normally involves collaborations between individuals within and among research groups (Chap. 4). Your mentees need to learn what is expected in such collaborative groups. You may expect respectful interactions among members of your research group and open sharing of ideas and data, but these behaviors and actions may be unfamiliar to, and need to be learned by, undergraduates new to research.

5.1.2   The Importance of Communication

The process of setting and meeting expectations is iterative and requires clear, open communication between you and your mentees. Communication occurs in many ways and on many levels. Many research labs have developed formal codes of conduct that lay out basic expectations for every member of the research team. In these documents, expectations are communicated as formal written or spoken statements, such as "Every member of the research team should be in the lab from 8 a.m. until 5 p.m." or "All the work done in this research group will meet the highest ethical standards for research."

You should have specific expectations for undergraduates in addition to the general expectations that apply to every member of the research group. These specific expectations might include that each student learns and can articulate the broader context of the research in a written proposal or a short talk, or prepares a poster on their research to present in a student symposium or regional meeting. Successful undergraduate researchers also have their own set of expectations. As a mentor you should know what their expectations are. Discussing these expectations with your mentees can provide insights into their understanding of research and what they want to achieve from their undergraduate research experience. These discussions will also help you to refine your own expectations and work effectively with your mentees to realize their expectations.

But actions matter more than words, and the informal ways we communicate our expectations through our actions and behavior can be more important than any formal document. For example, you may expect everyone to follow all safety requirements while working in the lab. If these requirements include wearing long pants, a lab coat, and close-toed shoes in the lab but you walk into the lab in shorts and sandals, you cannot expect that your mentees will pay much attention to this (or any other) expectation you've presented.

5.2 Empowering Your Students

At the beginning of this book, we defined a mentor as an individual who sees potential in their mentees and empowers them to become self-aware, independent-minded scientists and scientifically literate members of their communities (Chap. 1). In the short amount of time an undergraduate has for their research experience, you want to help your mentees see themselves as "real" scientists. Because we know that being a scientist involves constant learning and intellectual, personal, and professional growth, a successful undergraduate research experience should put your mentees on this same road. Getting on this road need not require publishing a *Science* paper based on a 6-week research experience; it is enough that the student finishes their research experience by arriving at a personal or intellectual level that is further along than where they were when they started, and that they are ready to continue on. Empower them to take those steps.

5.2.1 Ceding Some Intellectual Ownership

Even though most undergraduate research experiences are relatively short (6–10 weeks in a summer session, perhaps a semester or an entire year as part of their regular undergraduate curriculum), one of the most powerful ways to launch a student towards a STEMM career is to let them "own" their research. There are many ways to give students intellectual ownership of a research project. Undergraduates who have had some other research experiences may already have a project in mind. If it's closely related to your own research program, work with them to ensure that their results will tie in, amplify, or support it in ways that can advance it, perhaps in new directions (see Vignette 2.1 in Chap. 2). If it's unrelated to your own research program but you are interested in the topic or idea, and if you can support it and them through the process, let them do it. But if it's unrelated to your own research program and you're uninterested in the topic or don't have the time to support it or them, be honest about that and direct them to someone who can support them.

For undergraduates who had little or no previous research experience, you should identify a research question that you're interested in enough so that you'll be happy discussing it with your mentee and helping them see it through. It could be a pilot for something you'd like to develop into a future grant proposal, an exploration of an idea you've thought would be interesting but never had time to pursue, or a defined part of a larger, already funded research program. Alternatively, you could describe the broader context of your own research program and ask your mentee to think about some specific topics within that context that they might find interesting to pursue.

Regardless of how much previous research experience your mentees have, whatever research project they're pursuing should have a good probability of "success." One way to increase the odds of success is to be clear from the outset about what defines success (see Chap. 7). Is it a statistically significant result, learning the process of research, or something in between?[3] Be aware that what success in a research project means to you as a more senior researcher is not likely to be the same as what success in a research experience means for an undergraduate researcher.

5.2.2 The Importance of Regular Feedback

Feedback plays a crucial instrumental role in the development of any researcher, but it is especially important for new researchers. Feedback helps develop independence in your mentees and empowers them to push beyond their initial limits and deal with unexpected outcomes. Constructive and supportive feedback provides your mentees with a "reality check" on how their research is going and their progress towards their own goals in STEMM. At every step along the way, feedback helps students develop independence of intellectual thought and ideas. As a mentor, providing regular feedback can help your mentees re-calibrate their ideas, adjust their thought processes, and encourage them to keep going with the day-to-day, often tedious, tasks of research.

We hope that all our research projects work as planned and that our mentees achieve all of their goals and expectations. But sometimes things don't go as expected (Vignette 5.1). Those times can be the most important part of a research experience. How do you help your mentees feel empowered to try again and move forward when "everything went wrong"? How do you say "it's gonna be okay"? There's no uniform answer, but when things don't go as expected it's crucial that you provide rapid supportive and constructive feedback and help refine or redirect your mentees' expectations into something achievable based on your own experience. This is also an opportunity for you to ask for feedback from your mentees about matches and mismatches between your expectations and theirs.

Vignette 5.1 This time it didn't work: remote mentorship and the magnetic field of a tractor

Contributed by Rick Saltus

The 2020 RECCS was done 100% virtually because of COVID-19. Mentors and mentees intended to "meet" online several times a week. Our project is CrowdMag—we use smartphones to collect data to learn about our magnetic environment. For the previous 3 years we have mentored students in person on field data collection and computer work. The students present their research as scientific talks and posters. Based their research, the previous years' mentees helped develop the 2020 experiment.

In the 2020 virtual RECCS, we made an initial connection with a student, learned about his background in rural southern Colorado, and discussed potential magnetic targets in his local environment. We proposed, "How about measuring the magnetic aura of a big tractor?" He replied, "Yes, sounds good." We sensed a connection and enthusiasm. Difficulties on his side included access to a working phone, so we bought one and mailed it to him. He collected a few data points, and then communication between us ceased. But, we didn't give up on him. We used the data he collected and made a poster for him to present at the final session. But he didn't show up.

It was a frustrating situation for sure, but I hope that lessons were learned or seeds were sown that may blossom down the road.

5.3 Future Research with and by Your Mentees

A short undergraduate research project can propel your research forward or send you and your mentees in new research directions. It can also help guide your mentees into a STEMM career. In mentoring undergraduate researchers, you should be paying attention to all of these opportunities. You should be open to exploring new directions identified with your mentees. If they are interested in continuing the research beyond the time defined by their research experience, think about how that could work. Do you have time, energy, or inclination to continue working with them? Or would it be better to let them continue that direction working with others in your (or another) research group, or on their own? How can you facilitate these different directions? Does the student have other skills and interests that you can help them develop either to advance your own research or in other STEMM fields? Being open to asking these questions of yourself and your mentees is a key part of being a mentor. We explore in more detail some strategies for asking and answering them in Chap. 8.

5.4 From Mentors to Program Directors: The Science Behind Successful Undergraduate Research Programs

Many undergraduate research experiences happen within larger undergraduate research programs. These are usually directed by a senior researcher who has identified a broad research theme or question amenable to undergraduate research and recruited a number of mentors to work with the undergraduates on various parts of it. Directing an undergraduate research program comes with additional responsibilities. The program director needs to secure funding in addition to that for their own research program (see Sect. 2.5 in Chap. 2), sets overall recruiting goals (see Sect. 3.6 in Chap. 3), works with the other research mentors to articulate the overall code of conduct for all of the participants—mentees *and* mentors—in the research program, has additional oversight and reporting requirements, and may take on other administrative responsibilities. Although it may seem like there's little time left to do your own research, that need not be the case.

At first glance, you might think that as a program director that you are only facilitating research by others and that the undergraduate research program itself cannot provide you with additional research opportunities. If others can help you with some of the day-to-day operations, you may be able to have your own slice of the undergraduate research pie and work with one or more mentees on specific research projects. As a program director, you may be able to budget for additional staff who can support various aspects of the undergraduate research program. At some large universities, many of the administrative and logistical tasks are handled by professionals in an office for undergraduate research. If you have a strong, supportive team handling the logistical details, you may have the opportunity to be a research mentor in a broader research program that you also direct. In this role, you should define and model the "best practices" of mentorship that you expect from the other mentors in the program.

As the program director, you also set the broad research theme of the undergraduate research program. Use this theme to recruit mentors who, together with you, can form one or more research teams working together to address more challenging questions. As professional scientists, we work in collaborative teams. The most successful undergraduate research programs involve teams of mentors and undergraduates. Each member of the team brings unique expertise and skills to the project and "owns" part of it, but works together with the others on research into complex problems.

And then there is the science of designing, running, and evaluating successful undergraduate research programs. Most STEMM researchers have little or no experience in formal evaluation of the different aspects of research programs. Formal study of your research program could lead you to identify new collaborators (perhaps in an Education or Sociology Department in your own institution or one nearby). Consider recruiting an undergraduate researcher to work with you on evaluating and improving the research program itself. Working with new colleagues and mentoring students outside your own field will introduce you to new ideas and skills, and could lead you into new research directions. We discuss how to evaluate undergraduate student research experiences and programs in more detail in Chap. 7.

5.5 Take-Home Messages

✔ Setting expectations and open communication is a two-way street.
✔ Getting research done comes first but ceding some of your control to your students allows them to "own" their research.
✔ Feedback and support is crucial when unexpected outcomes or roadblocks arise.

Chapter 6
STEMM Education is More Than Training

Abstract Successful research experiences are not lab exercises with known outcomes or inquiry-based learning activities. Rather, undergraduate research experiences are collaborations between mentors and mentees in which the key benefit to the mentee is learning how research gets done. This chapter outlines ways for mentors to effectively teach the practice of research so that mentees are active learners. A key point of emphasis is the difference between training and learning: training is the inculcation of rote habit whereas learning is what can happen throughout life for those willing to risk it. Mentors are teaching students to be researchers, not training them. And mentees are learning from their mentors the practice of doing research.

Successful research experiences are not lab exercises with known outcomes or inquiry-based learning activities. Rather, undergraduate research experiences are collaborations between mentors and mentees in which the key benefit to the mentee is learning how research gets done. In this chapter, we outline ways for mentors to effectively teach the practice of research. We emphasize here that as a mentor, you are not training your students to be researchers. Rather, your mentees are learning from you the practice of research. What's the difference? Training is the inculcation of rote habit, and is how one instructs an animal. In contrast, Learning is what can happen throughout life for those willing to risk it (Orr, 1992, pp. x–xi).

Training is about imparting specific, often technical, skills that have immediate practical applications. After training, a student may be able to do lab or field work better or more efficiently, but they haven't really learned how to do research. In contrast, education is based on concepts and seeing the bigger picture. Through their education, mentees should develop a sense of the general concepts of how to do research and your research program. They should be able to think through the implications of what they're doing and learning. Your mentees do need training in responsible conduct of research, lab or field safety and research operation, and particular technical skills; the one Vignette in this chapter (Vignette 6.1) illustrates the importance of training. However, in an undergraduate research experience, training is only the start of truly learning about the research process. The latter is what distinguishes what a mentee gets out of an undergraduate research experience from what they learn in nearly every science class.

© The Author(s), under exclusive license to Springer Nature Switzerland AG 2022   55
A. M. Ellison and M. V. Patel, *Success in Mentoring Your Student Researchers*,
SpringerBriefs in Education,
https://doi.org/10.1007/978-3-031-06645-0_6

6.1   Basic Training

6.1.1   Ethical and Responsible Conduct of Research

The responsible conduct of research ("RCR") is the most essential part of the scientific enterprise. But what is RCR, why is it important, and how do you go about inculcating it in your mentees?

RCR is the umbrella term for promoting the goals and objectives of scientific inquiry, and creating and nurturing a collegial and inclusive environment that allows diverse scientists to work together toward agreed-upon, common goals. Practiced regularly and seriously, RCR results in scientific data, results, and outcomes that are trustworthy.[1] We used to teach our mentees about RCR by off-handedly mentioning particularly egregious occurrences of another scientist's irresponsible, unethical, or illegal conduct, such as data fabrication or falsification, plagiarism, financial mismanagement of grant funds, or harassment. We would self-righteously assert that we would never do such a thing, and by implication, urging our mentees not to do such a thing. Now, however, scientists at every career stage, including undergraduates, are required or strongly encouraged to take RCR training sessions or workshops before they start to do research. Scientists are also expected to take refresher training courses in RCR at regular intervals (at least every year or two) to stay up to date.

RCR workshops emphasize positive behaviors while illustrating negative behaviors. Common goals for RCR include:[2]

- Developing and nurturing a culture of integrity;
- Empowering researchers to hold themselves and others to high ethical standards while discouraging and preventing unethical conduct;
- Increasing knowledge of, and sensitivity to, ethical issues in the conduct of research by scientists from diverse backgrounds and cultures;
- Increasing the appreciation for what are acceptable and ethical scientific practices, and what are the regulations, policies, statutes, and guidelines governing the conduct of research in one's home institution or country and in other institutions or countries;
- Improving the ability to make responsible choices when faced with ethical dilemmas in research practice.

The subject matter of RCR workshops begins with general topics applicable to all STEMM fields. These overview workshops usually take several hours and include topics such as:

- Responsibilities and relationships between mentors and mentees;
- Civil behavior, including harassment, bullying, and other inappropriate behaviors;[3]
- Definitions and examples of research misconduct and questionable research practices;
- Data management, including data collection, keeping good, written (electronic) records, data ownership, data archiving, and data sharing;[4]

- Issues around publications (McNutt et al., 2018), including guidelines for (co)-authorship, peer-review, copyright, and open-access;[5]
- Identification of conflicts-of-interest, which can include financial conflicts and interpersonal conflicts;
- Overviews of issues regarding safe operations in the lab or field and doing research on human or animal subjects;
- The relationships between scientists, the environment, and broader societies.

6.1.2 Doing Research Safely

As a professional scientist, you are expected to know the risks and hazards involved in your research and to be able to work in an environment that reduces your risks and hazards to the extent possible. You have likely taken training courses and regular refreshers that cover the rules, regulations, and safe practices required to handle and work with hazardous equipment or materials. Your mentees need the same training courses to learn the basic rules and skills of conducting safe and responsible research (Vignette 6.1). Training courses should be part of the orientation or onboarding of your mentees. They also need to see that you, their mentor, takes these training courses seriously. You should emphasize the importance of these courses rather than denigrate them as "something we all just have to get through." If the trainings are open to both mentors and mentees, attend them together. If not, follow up with your mentees to find out what they learned and how they—and you—are going to maintain safe and responsible research practices.

Vignette 6.1 Formal training helps you and your students mitigate bad outcomes

Contributed by Allyson Degrassi

Attending a training session is one thing. Paying attention during training sessions is another. I was taking a training course with my students on how to safely and ethically handle live animals when the instructor clearly said, "Make sure your surroundings are clear before you attempt to remove your animal from the trap to prevent injury." Not a minute passed before a student, who had NOT been paying attention to the training, tried to clear a trap while standing in front of a boulder. The student swung the trap with the animal inside and whacked it on the rock, killing the animal instantly. The instructor rushed to the animal and tried to revive it. Upon seeing that the animal could not be revived, the student began to cry. As required by IACUC protocols, the instructor reported the incident, which led to the student being temporarily banned from doing further research with animals. As if the unnecessary loss

> of a life was not enough, this also could have resulted in a halt to the experiment. I tell this story to all of my student researchers and ask them the question, "Was this an accident or negligence?" The correct answer is "negligence" because the student was willfully not paying attention.

Model good behavior yourself. You want to be respected for who you are and what you can do, and you need to be respectful of the students, staff, and other researchers in your research group. Completing regular training courses in recognizing and preventing harassment based on gender, age, ethnicity, etc. are expected in any institution. Acknowledge when your mentees *and* your more senior colleagues work safely and responsibly and call them out when they don't.

6.1.3 Teaching Technical Skills

Your mentees will certainly need to know specific technical skills to do research with you. These can range from simple skills, such as how to use a pipette or a microscope, to more complicated ones such as data organization and statistical analysis. Learning technical skills allows your mentees to get started and move the research along while helping them gain confidence in their ability to be a scientist and do research.

How you teach technical skills is the key to effectively integrating them into a research experience. You want to strike a balance between demonstrating or describing skills and allowing the student to learn and develop them independently. You want to spend the time needed to ensure that the mentees can not only use their newfound skill set to perform required tasks well, but also explain why they are done the way they are done. It is rare that anyone just beginning to do research picks up a new skill on their first try. You should provide plenty of opportunities for your mentees to practice skills, get them "wrong," ask questions about them, and suggest new ways of doing them that you may not have thought of but may be easier or more efficient. Even if your more experienced mentees are certain that they know how to do a certain task, you still want to take the time to show them. There are likely to be differences among research groups in techniques, protocols, and equipment (e.g., not all pipettes are the same).

6.2 Teaching the Research Process

Your mentees may have learned *about* the research process in a classroom. But teaching about the research process in a classroom is very different from teaching the process of research through an undergraduate research experience. The former is primarily a theoretical exercise whereas the latter is learning through doing. Consider the difference between asking a student to read and then summarize an article on constructing and distinguishing among multiple working hypothesis (e.g., Chamberlin, 1890; Platt, 1964; Elliott and Brook, 2007) *versus* actually coming up with a research question and even a single testable hypothesis given the overall context or theme of your research group. Be a role model: discuss your own struggles with formulating research questions and hypotheses.

6.2.1 Experience Matters

It's important to remember that your mentees are in an undergraduate research experience to learn how to do research, but they are not graduate students or post-docs. They likely have little or no previous experience working as a member of a research group and they're learning norms of behavior and interactions (Chap. 4) at the same time that they're learning about the research context and themes (Chap. 5). Your mentee may ask what you now consider to be a "trivial" or "straightforward" question or hypothesis, but you've got many more years of experience thinking about the system and they are just starting out. As with teaching research skills, you want to provide time and space for your mentees to try out ideas, ask questions, and find their own answers. As a mentor, you should be open to even the "simplest" of questions. On reflection, you may even be surprised how often simple questions lead you to complex, unexpected answers.

6.2.2 Teaching Without Giving Grades

Another difference between teaching about the research process to a class of students and teaching the research process by doing research with your mentees is that the undergraduate research experience is not graded. Your mentees' stipends are not contingent on performance (see Sect. 2.2 in Chap. 2) and there is no report card issued at the end of their experience. Rather than constantly evaluating your mentee against an abstract standard, you should work with them as an individual to further themselves along their own pathway in STEMM (see Sect. 5.2 in Chap. 5).

There are two keys to successfully mentoring each individual student in the process of research: understanding their capacity; and adjusting your expectations and teaching style to match their abilities and modes of learning.

Understanding Your Mentee's Capacity

Every student in your research group is an individual and each has a different capacity to learn the process of and do research. Their capacity depends on many factors, including their cultural backgrounds, level of education, openness to new ideas, and general enthusiasm. All of these factors can and do change over time.

Be Flexible Yourself

We all have preconceived notions and biases about what an undergraduate researcher should be like. These qualities were expressed, either implicitly or explicitly, in how you designed and funded an undergraduate research opportunity (Chaps. 1 and 2). They were in your mind when you were recruiting students, reading their applications and letters of recommendation, and interviewing them (Chap. 3). The real student now working with you may meet all your expectations, but they also may not. You may find that you need to spend more time on teaching technical skills that you expected or your mentee may be more prepared to creatively design experiments than you thought. As a mentor, you need to understand and empathize with your mentees as individuals, quickly learn their strengths and weaknesses, and take advantage of opportunities you see to catalyze the necessary changes to build their capacity.

6.3   Professional Development

Somewhere in between basic training of technical skills and teaching the research process, you need to work with your mentee to develop the many other skills that will help them develop as a professional scientist. Think about all the things you do on a regular basis that contribute to your success in STEMM. For example, you probably:

- write, edit, or review proposals and papers;
- write and send out letters of recommendation;
- teach classes;
- give talks of different lengths and scopes to audiences of specialists and nonspecialists;
- find and apply (or have applied) for jobs in the field;
- hire, supervise, and evaluate staff;
- recruit and mentor students;
- develop and work with teams of researchers.

Where did you learn how to do all these things? Do you do some of them well and ignore the ones you don't or dislike? Many of us will say that we picked up these skills when we needed them or mimicked the ways our advisors and mentors did them. Perhaps we were in a research group or lab meeting where these kinds of skills were discussed. And a few of us may have attended workshops or short-courses focused on one of these skills.

But unlike responsible conduct of research and technical skills, you may not think that you are the best person to teach your mentees all these skills. You may have facility in one or more of them. If, for example, if you are a journal editor, you may be in a good position to teach about how to write an effective abstract or a paper. But just because you took a short-course in personnel management or another professional skill doesn't mean you can teach it effectively to others. Rather than trying to teach a wide range of professional development skills to your mentees, help and encourage them to seek out workshops and short-courses on your campus or at a nearby one.

If their research experience is part of a larger undergraduate research program, the program itself may offer workshops in some of these topics and expect all the students to attend (see Sect. 6.4, below). Be supportive and discuss how important it is to learn and continue to improve these skills. Allow them the time to take these courses and, if necessary, cover the costs associated with them. Be a role model: if the opportunity arises for you to improve your own skills through one of these professional development workshops or short-courses, attend it with your mentee. Then, complete the circle by bringing the ideas and tools back to your larger research group as part of a regular lab meeting or in another organized context.

6.4   From Mentors to Program Directors: Teaching Skills Needed by STEMM Professionals

The larger number of students and mentors working and learning together in undergraduate research programs creates an "economy of scale" in which a wide range of skills can be taught effectively and efficiently. Training courses that cover basic requirements of doing work in STEMM, such as responsible conduct of research, lab or field safety, and ensuring a safe and respectful working environment are offered on most college or university campuses, in governmental agencies, and in the private sector. As you organize your undergraduate research program and provide orientation materials to the students and mentors, you can ensure that everyone takes these training sessions at the beginning of the program. Be sure to budget for trainers and facilitators and get the sessions scheduled well in advance.

In a similar vein, the students in an undergraduate research program will benefit greatly from interactive workshops in which they learn and practice professional skills. Such workshops can be facilitated by individual mentors in your research program who have the necessary expertise and are interested in teaching the skills or by professionals (e.g., individuals from an on-campus writing center, journal editors, human-resource staff). If these workshops are limited to the students themselves (i.e., without mentors in the room) they can provide safe spaces for students to try out new skills, and fail at them repeatedly before they finally "get it right." Such workshops also can provide opportunities for students with different abilities to develop as peer mentors.[6] And as with the training sessions, be sure to budget for the necessary facilitators, materials, and supplies, and get them on the calendar so that mentors know when their mentees will be participating in the workshops.

Finally, research mentorship itself is a skill that can be learned and improved over time. As a program director, you should be a role model for your team of mentors. You should also provide time and support for them to participate in workshops and courses in mentoring, either at your home institution or at various mentoring institutes around the world.

6.5   Take-Home Messages

✔ Teach research by doing research.
✔ Be a role model for your mentees as they learn different aspects of the research process.
✔ Professional development skills, including mentorship itself, can be learned and practiced in experiential workshops.

References

Chamberlin, T. C. (1890). The method of multiple working hypotheses. *Science, 15,* 92–96.
Elliott, L. P., & Brook, B. W. (2007). Revisiting Chamberlin: Multiple working hypotheses for the 21st century. *BioScience, 57*(7), 608–614.
McNutt, M. K., Bradford, M., Drazen, J. M., Hanson, B., Howard, B., Jamieson, K. H., et al. (2018). Transparency in authors' contributions and responsibilities to promote integrity in scientific publication. *Proceedings of the National Academy of Sciences, 115*(11), 2557–2560.
Orr, D. W. (1992). *Ecological literacy: Education and the transition to a postmodern world.* Albany: State University of New York Press.
Platt, J. R. (1964). Strong inference. *Science, 146,* 347–353.

Part III
Mentoring Beyond the Research Experience

The final section of this book looks at mentoring beyond an individual research experience. In the first chapter of this section, we present various ways to evaluate the student and their research, as well as the mentor's activities. In the following chapter, we outline ways to continue to support students to continue their research beyond the formal time frame of their individual research experience. In the final chapter of the book, we discuss fledging mentees and long-term continuation of the mentor-mentee relationship.

Part III
Merging Beyond the Research Enterprise

Chapter 7
Evaluation

Abstract Successful mentoring of undergraduate researchers doesn't end with the conclusion of a relatively short research experience. Taking the time to evaluate the research experience—its short-term effects on the students and on yourself as a mentor—is invaluable. This chapter presents a number of ways to evaluate students' success in their research and your own mentorship. Many of these methods—grounded in educational theories—likely will be unfamiliar to STEMM researchers: "experiments" may seem unstructured, unplanned, or poorly designed compared with more familiar lab and field experiments; "control" groups usually are lacking; and the "data" are more likely to be verbal and qualitative than numeric and quantitative. Any evaluation should be considered as an adaptive mechanism to continually refine and improve undergraduate research experiences.

Successful mentoring of undergraduate researchers doesn't end with the conclusion of a relatively short research experience. Taking the time to evaluate the research experience—its short-term effects on the students and on yourself as a mentor—is invaluable. You should document and share with your fellow mentors your successes, failures, and lessons learned so as to improve the range of ways the entire community can foster each new cohort and generation of STEMM researchers. And each time you work with one or more undergraduate researchers, you have new experiences yourself and you learn things that help you be a better mentor and researcher in the future.

In this chapter, we discuss the many ways you can evaluate your students' success in their research and your own mentorship. Many of these methods will be unfamiliar to STEMM researchers: "experiments" may seem unstructured, unplanned, or poorly designed compared with more familiar lab and field experiments; "control" groups usually are lacking; and the "data" are more likely to be verbal and qualitative than numeric and quantitative. Nonetheless, there is a range of established methods and a rich literature that describes the analysis of both formal and informal methods that can be used for student and mentor self-reflection and for giving and receiving feedback to and from students and mentors.

We encourage mentors and program directors to partner with researchers in education departments to ground their evaluations in educational theories. There are also

A. M. Ellison and M. V. Patel, *Success in Mentoring Your Student Researchers*,
SpringerBriefs in Education,
https://doi.org/10.1007/978-3-031-06645-0_7

professional evaluators who can help develop, test, validate, and analyze the resulting data. But whether you do the evaluations yourself or in partnership with others, we encourage you to treat any evaluation as an adaptive mechanism to continually refine and improve undergraduate research experiences.

7.1 Why Evaluate?

An undergraduate research experience is widely considered to be one of the most impactful and formative events in the life of a professional STEMM researcher (Linn et al., 2015; National Academies of Sciences, Engineering, and Medicine, 2017; Ma et al., 2020). If this is so well-known, why do you need to evaluate your students and your work with them? First, every student research experience is different and simply engaging a student in research doesn't guarantee its short- or long-term success. You should want to know that *your* experience of doing research with students actually had a positive impact on them. Second, you need to consider different dimensions of "success" and "impact." On the one hand, you should have discussed success and impact with your students at the outset of their research experience (see Sect. 5.1 in Chap. 5). On the other, there may be broader expectations or definitions of success in undergraduate research, or legal requirements to track certain outcomes of undergraduate research experiences.[1]

Third, we evaluate almost everything else we do in STEMM: our classes, experimental methods and results, clinical outcomes, and employment (e.g., performance reviews, promotion, tenure packages). Why shouldn't we evaluate how we do as mentors? You probably didn't decide to mentor only one student or a single cohort of undergraduate researchers. Rather, you see mentoring as part of your responsibility as a STEMM researcher and you are in it for the long haul. Periodic evaluation of your own mentorship skills will help you continually improve them and make the undergraduate research experience productive and positive for both yourself and your students.

7.2 Evaluation in Practice, in Theory, and in Practice Grounded in Theory

Evaluations and assessments of undergraduate research experiences have been done almost entirely on cohorts of students participating in undergraduate research programs. Evaluation and regular reporting on outcomes of undergraduate research programs now is commonly required (or at least incentivized) by the funding agencies that support the programs. Thus, our discussion focuses on these types of evaluations and assessments, but there is no reason that these methods can't be modified and applied by you if you are a solo mentor regularly working with one or two undergraduate research students.[2]

7.2.1 Are You Evaluating for Yourself or For Others?

The first and perhaps most important question to ask is: What is your primary goal in evaluating and assessing your own students (in your undergraduate research program or your own research group)? Is it to determine whether you and your students successfully accomplished your goals and can use the information to do better the next time? Or, is to provide local data that can be incorporated into broader (e.g., national) assessments that evaluate goals and measures of success for undergraduate research programs? These are very different goals, and even identical questions may provide answers or data that have different meanings in the different contexts (e.g., local *versus* national, groups of small undergraduate institutions *versus* groups of large research universities).

7.2.2 The Value of Individual Case Studies

Most undergraduate research mentors and program directors are interested primarily in improving the effectiveness of their own research group or local program. If you're interested in putting your own results into a broader context, chances are it's to show on your next grant application that you're doing better than the competition in fostering undergraduate researchers. Even though evaluating yourself or your program is a laudable goal, it is more complicated than you might think.

At least through 2010, if individual undergraduate research programs were evaluated and assessed at all, program directors and their collaborators developed or selected their own protocols or survey instruments, individually implemented or administered them, and analyzed their own data. These surveys usually consisted of student self-assessments of learning gains or other outcomes, and the students completed "summative" surveys (sensu Plotner, 2018) only at the end of their research experiences.[3] Qualitative data from these *post-hoc*, self-assessment surveys certainly have provided some insights about student gains following research experiences. In general, students from minoritized groups show the greatest increase in learning gains (e.g., Lopatto, 2004, 2007; McDevitt et al., 2016). However, the data collected in these and other similar studies were from neither a random nor a representative sample. When they were shared with others (and many were not), these individual programmatic assessments were presented as unique case studies with limited broader applicability (e.g., Seymour et al., 2004; Dávila et al., 2013; McDevitt et al., 2016).

Quantitative evaluations of undergraduate research programs are uncommon. Most research groups and programs now collect basic demographic data on student participants to demonstrate commitments to broader goals of increasing diversity in STEMM, but diversity and demographic data are very weak measures of success. Some programs have used "off-the-shelf" quantitative surveys, but these usually consisted of conceptually ambiguous questions that were rarely validated and were not comparable among programs. Nonetheless, both these qualitative and quantitative surveys have helped individuals and programs demonstrate their effectiveness and improve themselves (Linn et al., 2015; McDevitt et al., 2016).

7.2.3 National-Level, Cross-Program Surveys

It is perhaps not surprising that, even as we completed this book in 2022, there were few standard survey instruments for evaluating the effectiveness and impacts of undergraduate research. The flexibility in design and implementation of many undergraduate research programs afforded by granting agencies such as the US NSF or NIH encourages innovative pedagogical approaches but also increases heterogeneity among programs.[4]

One of the first broad-based assessments of undergraduate STEMM research programs used the Survey of Undergraduate Research Experiences (SURE) (Lopatto, 2004). This survey was semi-quantitative (most questions had answers that could be coded on an ordered [Likert] scale) and was funded by the US-based Howard Hughes Medical Institute.[5] The SURE survey was administered to students immediately upon their completion of their undergraduate research experience; there were no "pre/post" comparisons. The results supported broad expectations for undergraduate research experiences: participants—especially female and minoritized students[6]—were generally pleased with their experience and reported gains in learning about the research process, scientific problem-solving, lab skills, and personal development (including willingness to overcome obstacles and ability to work independently). Less than 5% of the >1000 students completing the survey no longer planned to continue their studies in STEMM fields (Lopatto, 2004).

A comprehensive assessment tool for NSF-supported REU programs in biology— the Undergraduate Research Student Self-Assessment (URSSA) survey—was implemented in 2010 (Hunter et al., 2009).[7] The standard implementation of URSSA provided data to NSF on how well REU programs met national programmatic benchmarks. URSSA provided comparable data and results that were qualitatively similar to those obtained from SURE. Like SURE, URSSA was normally given to students only at the end of their research experience and could not measure changes in student learning or other programmatic goals as a result of their participation in an undergraduate research experience. Although it is still freely available to researchers and undergraduate program directors, URSSA was superseded (in 2019) by CIMER (the assessment tool of the Center for Improvement of Mentored Experiences in Research).[8] Most NSF-REU grantees are now strongly encouraged or required to use the CIMER assessment tool to evaluate their undergraduate research programs.

7.2.4 The Importance of Theory

Individual site-based surveys and cross-site surveys like SURE, URSSA, and CIMER serve their intended purpose, but all of them lack theoretical underpinnings that make it difficult to relate their findings to the broader literature on education and development of student learning skills or to understand similarities and differences among experiential undergraduate research programs (e.g., Beninson et al., 2011; Linn et al., 2015; Wilson et al., 2018). For example, we previously analyzed 10 years of before/after ("pre/post") surveys of student participants in the Harvard Forest Summer Research Program in Ecology (McDevitt et al., 2016).[9] We designed, tested, validated, and deployed a short self-reporting survey that intentionally balanced sample size and survey depth.[10] In our survey, we asked typical questions about topics scientists like ourselves think are important and interesting (e.g., improvement in ability to do lab or field work, present scientific data, work independently or in research teams) rather than those that educators might have identified as central to learning science (we discuss those below, in Sect. 7.3).

Continued use of tools used to evaluate undergraduate research experiences that are not grounded in educational theories may let us know that undergraduate research experiences *are* valuable and successful to *our* students, but are unlikely to tell us *why* they are successful for them or for larger populations of undergraduates. Yet, knowing why they are successful (or not) is critical to improving undergraduate research experiences.

7.3 Theories and Resources for Evaluating Undergraduate Research Programs

7.3.1 Educational Theories for Evaluation

As we have stressed throughout this book, undergraduate research experiences are more than just doing research. They also are crucial for developing a student's identity as a scientist and for building capacity in STEMM. Undergraduate research experiences also connect students to broader personal and professional networks of peers that will nurture and sustain them for many years. Pathways to success differ widely among individual students, cohorts of student researchers, and undergraduate research programs. Contexts—a student's background, interactions within and between cohorts, settings of undergraduate research programs, etc.—are crucial but often underappreciated reasons for the success or failure of students, cohorts, and programs. Adaptive change in mentoring strategies and transfer of "best practices" from "model" programs to other undergraduate research programs also depend strongly on contextual similarities and differences among programs. These contextual differences are not captured in self-assessment surveys of gains in learning, skills, and confidence in doing STEMM research.

How do students learn? Many theories of learning recognize and incorporate contextual differences and social and cultural influences. Common to most of these theories is the idea that learning is culturally mediated: words, texts, social cues, and other symbolic objects fundamentally shape how an individual constructs knowledge.

There are many sociocultural learning theories, but most include at least one of three themes (Vygotsky, 1980; Wertsch, 1993):

1. Learning is less about accumulation of knowledge than performance in different social contexts (theories of **situated cognition** [Brown et al., 1989] and **embodied cognition** [Wilson, 2002]).
2. Knowledge is co-constructed with other individuals or by using psychological tools (theories of **situated learning** [Lave and Wenger, 1991], **distributed intelligence** [Pea, 1993], **socially shared cognition** [Resnick et al., 1991], and **distributed cognition** [Salomon, 1997]).
3. The environment, community, or culture shapes how an individual learns (the **bioecological theory of human development** [Bronfenbrenner and Morris, 2006], the theories of **cultural psychology** [Cole, 1998] and **cultural learning** [Tomasello et al., 1993], **activity theory** [Engeström et al., 1999], and **cultural historical activity theory** [Roth and Lee, 2007]).

One theory that includes all three themes and that flexibly accommodates nearly all of the concepts proposed in the other sociocultural learning theories is cultural-historical activity theory (CHAT) (for a detailed exposition of CHAT, see Roth and Lee, 2007; Yamagata-Lynch, 2010). We have applied CHAT successfully in evaluating and assessing the Harvard Forest Summer Research Program. McDevitt et al. (2020) provide a framework and template questions that other undergraduate research programs in STEMM could use to evaluate their programs using CHAT.

7.3.2 Partnering for Success

As a research scientist, a teacher, a mentor, and perhaps a director of an undergraduate research program, department chair, or dean, your plate is already overfull. You will not be motivated to learn how to evaluate your research with undergraduates unless you can see immediate gain from the evaluation—for yourself, your students, or your research and teaching program. The good news is you don't have to learn another discipline or become a professional evaluator to ground your evaluations in theoretical contexts.

If you want an intellectual partner with whom to work on evaluation and assessment of undergraduate research, look in your campus directory for colleagues with the skills you want to learn or wish you had. Walk across campus to the Department or School of Education or Psychology. Talk to the graduate students or faculty. Attend their seminars. Listen and learn, seek out allies, build connections, write proposals. On the other hand, if you just want to "get it done" without any (or with very little) intellectual commitment, budget for and hire a professional evaluator.

7.4 Learning from Evaluations and Evolving Undergraduate Research Experiences

In our own research, our pet hypotheses aren't always right, our experiments don't always work, and our papers are reviewed, critiqued, and undergo substantial revision before being published. A similar and continually iterative process of trial-and-error, assessment, revision, and re-implementation is essential for successful mentoring of undergraduate STEMM researchers and running successful experiential undergraduate research programs in STEMM. Just like you use your data and analyses to test scientific hypotheses and refine your understanding of your research system, you can use data from evaluation and assessment to refine how you mentor undergraduate researchers and adapt your strategies to different individuals and contexts.

If evaluations are grounded in theoretical constructs and questions are framed as testable hypotheses, the results of evaluations of undergraduate research programs can be used to adaptively improve individual undergraduate research experiences and illuminate causes of successes—and failures—across undergraduate research programs in STEMM. It's important that you keep in mind that regular evaluation and change will not lead to a perfect program. You will always want to include elements in an undergraduate research program that not every mentor and mentee will appreciate or like. Despite your best efforts, not every student or mentor will succeed. In thinking about the importance of evaluation, always keep in mind Maya Angelou's dictum to "do the best you can until you know better. Then when you know better, do better."

7.5 From Program Directors to Mentors: Closing the Feedback Loop

Each of the other chapters in this book has concluded with how to translate individual mentoring activities to larger undergraduate research programs. But as we noted at the outset of this chapter, evaluations of undergraduate research experiences normally are done at the programmatic level, either by the program director, their colleagues, and their staff or by a professional evaluator and their team. To close the loop and continue to refine and improve your experiential undergraduate research program, you need to get the relevant feedback back to the individual mentors.[11]

Summaries of the evaluations should be distributed to all the individual research mentors as soon as possible—and well before it's time to organize the next round of the research program and re-enlist existing mentors or recruit new ones. Acknowledge outstanding performance and success, either publicly or privately. When opportunities for improvement are identified in the evaluation, work with individual mentors to implement appropriate changes. Find ways to reward improvement, but if individual mentors are unwilling to change, continue to underperform, or appear to be "burned out," encourage them to take a break from mentoring students in your program.

Always keep in mind that it takes a team working together and all pulling in the same direction to sustain an undergraduate research program. All team members share in the successes of the program and suffer its failures. As a program director and a mentor, you need to nurture the successes and prune the failures.

7.6   Take-Home Messages

✔ Use educational theories to understand why undergraduate research experiences are successful.
✔ Evaluate the research experience for yourself and your students.
✔ Use evaluations as an adaptive mechanism to improve mentoring and the under-graduate research experience.

References

Beninson, L. A., Koski, J., Villa, E., Faram, R., & O'Connor, S. E. (2011). Evaluation of the research experiences for undergraduates (REU) sites program. *Council on Undergraduate Research Quarterly, 32,* 43–49.

Bronfenbrenner, U., & Morris, P. (2006). The bioecological model of human development. In R. M. Lerner (Ed.), *Theoretical models of human development* (pp. 793–828). Hoboken: Wiley.

Brown, J. S., Collins, A., & Duguid, P. (1989). Situated cognition and the culture of learning. *Educational Researcher, 18,* 32–42.

Cole, M. (1998). *Cultural psychology: A once and future discipline.* Cambridge: Harvard University Press.

Dávila, S., Cesani, V. I., Medina-Borja, A. (2013). Measuring intercultural sensitivity: A case study of the REU program at UPRM. *Proceedings of the 120th Annual Conference and Exposition of the American Society for Engineering Education, ASEE,* Washington, DC, paper # 7538. http://www.asee.org/public/conferences/20/papers/7538/download.

Engeström, Y., Miettinen, R., & Punamäki, R. L. (Eds.). (1999). *Perspectives on activity theory.* Cambridge: Cambridge University Press.

Hunter, A. B., Weston, T. J., Laursen, S. L., & Thiry, H. (2009). URSSA: Evaluating student gains from undergraduate research in the sciences. *CUR Quarterly, 29,* 15–19.

Lave, J., & Wenger, E. (1991). *Situated learning: Legitimate peripheral participation.* Cambridge: Cambridge University Press.

Linn, M. C., Palmer, E., Baranger, A., Gerard, E., & Stone, E. (2015). Undergraduate research experiences: Impacts and opportunities. *Science,347*(6222), 1261757-1–1261757-6.

Lopatto, D. (2004). Survey of undergraduate research experiences (SURE): First findings. *Cell Biology Education, 3,* 270–277.

Lopatto, D. (2007). Undergraduate research experiences support science career decisions and active learning. *CBE-Life Sciences Education, 6,* 297–306.

Ma, Y., Mukherjee, S., & Uzzi, B. (2020). Mentorship and protégé success in stem fields. *Proceedings of the National Academy of Sciences, USA,117*(25), 14077–14083.

McDevitt, A. L., Patel, M. V., & Ellison, A. M. (2016). Insights into student gains from undergraduate research using pre/post assessments. *BioScience, 66,* 1070–1078.

McDevitt, A. L., Patel, M. V., & Ellison, A. M. (2020). Lessons and recommendations from three decades as an nsf reu site: A call for systems-based assessment. *Ecology & Evolution, 10*, 2710–2738.

National Academies of Sciences, Engineering, and Medicine. (2017). Undergraduate research experiences for STEM students: Successes, challenges, and opportunities. Washington, DC: The National Academies Press.

Pea, R. D. (1993). Practices of distributed intelligence and designs for education. *Distributed Cognitions: Psychological and Educational Considerations, 11*, 47–87.

Plotner, A. J. (2018). Summative evaluation. In B. B. Frey (Ed.), *The SAGE encyclopedia of educational research, measurement, and evaluation* (pp. 1636–1637). Thousand Oaks: SAGE Publications Inc.

Resnick, L. B., Levine, J. M., & Behrend, S. D. (Eds.). (1991). *Socially shared cognition*. Washington: American Psychological Association.

Roth, W. M., & Lee, Y. J. (2007). "Vygotsky's neglected legacy": Cultural historical activity theory. *Review of Educational Research, 77*, 186–232.

Salomon, G. (Ed.). (1997). *Distributed cognitions: Psychological and educational considerations*. Cambridge: Cambridge University Press.

Seymour, E., Hunter, A. B., Laursen, S. L., & DeAntoni, T. (2004). Establishing the benefits of research experiences for undergraduates in the sciences: First findings from a three-year study. *Science Education, 88*, 493–534.

Tomasello, M., Kruger, A. C., & Ratner, H. H. (1993). Cultural learning. *Behavioral and Brain Sciences, 16*, 495–511.

Vygotsky, L. S. (1980). *Mind in society: The development of higher psychological processes*. Cambridge: Harvard University Press.

Wertsch, J. V. (1993). *Voices of the mind: Sociocultural approach to mediated action*. Cambridge: Harvard University Press.

Wilson, M. (2002). Six views of embodied cognition. *Psychonomic Bulletin & Review, 9*, 625–636.

Wilson, A. E., Pollock, J. L., Billick, I., Domingo, C., Fernandez-Figueroa, E. G., Nagy, E. S., et al. (2018). Assessing science training programs: Structured undergraduate research programs make a difference. *BioScience, 68*, 529–534.

Yamagata-Lynch, L. C. (2010). *Activity systems analysis methods: Understanding complex learning environments*. LLC, New York: Springer Science+Business Media.

Chapter 8
Continuing the Research Experience

Abstract This chapter discusses ways to continue to support student mentees in research beyond the formal time frame of an undergraduate research experience. Both short- and longer-term avenues continuing to participate in research are presented. These avenues include students presenting mentored research at scientific meetings and conferences, expanding their research into undergraduate independent projects or senior theses, or taking the research in new, independent directions. Guidance is also provided on how mentors and program directors can sustain their enthusiasm for working with undergraduate researchers and larger undergraduate research programs.

In this chapter we discuss ways to continue to support your mentees in research beyond the formal time frame of their experience doing mentored research with you. We examine short- and longer-term avenues for them to continue to participate in research, including presenting their mentored research at scientific meetings and conferences, expanding their research into an undergraduate independent project or senior thesis, or taking their research in new, independent directions either in your research group or with another mentor. Continuing to work with your undergraduate mentees is easier if they are at the same institution or organization as you are, whereas additional challenges arise if they are at another one.

We also provide some guidance on sustaining your enthusiasm for mentoring undergraduate researchers and for managing and avoiding mentor "burn-out." In moving from individual mentors to undergraduate research program directors, additional considerations for sustainability include diversifying funding streams, getting buy-in from your senior administrators to contribute to long-term support for long-term undergraduate research programs, and building larger pools of diverse mentors. There is one Text Box and one Vignette in this chapter. Box 8.1 presents a series of questions to help you reflect on how well you met your goals as a mentor. Vignette 8.1 showcases an example of a new model for a large and sustainable undergraduate research program.

© The Author(s), under exclusive license to Springer Nature Switzerland AG 2022   75
A. M. Ellison and M. V. Patel, *Success in Mentoring Your Student Researchers*,
SpringerBriefs in Education,
https://doi.org/10.1007/978-3-031-06645-0_8

8.1   How Did It Go?

Before moving forward, give yourself some time to take stock of your experience as a research mentor. Inevitably, there were times you felt that the research was going well and that you were really connecting with your mentees. Other times, it was just a regular, uninspiring part of your "day job." And, there were probably a few times or incidents that stand out as really problematic or out-and-out failures. But most importantly, remember that your primary reason for being an undergraduate research mentor is to empower your mentees to become self-aware, independent-minded scientists and scientifically literate members of their communities. Did that happen? Are you ready to take on more undergraduate research mentees?

8.1.1   Take Time to Reflect

When you were thinking about becoming an undergraduate research mentor, you took time to consider whether, why, and how you wanted to mentor a student researcher (Box 1.1 in Chap. 1). You answered some questions about your own history as a mentee and what you got out of that. Before you jump right into more research with your mentee, we suggest that you should close the first loop by revisiting some of these questions. But don't look back at your earlier answers yet. Rather, answer the questions in Box 8.1 and then go back to your answers from Chap. 1. What do the similarities and differences in your answers tell you about your accomplishments in mentoring and what might be your next steps?

Box 8.1   Self-reflection: Did you accomplish your goals as a mentor?

Consider the experience you just had as a mentor. Ask yourself:

- How did I influence my mentees' development as scientists and peers?
- What barriers did I help them to overcome? If I couldn't help them directly, was I able to connect them with others of different genders, race or ethnicity, abilities, or economic backgrounds who could *and did* provide them with additional support or guidance?
- What did I learn from my mentees that contributed to making me a better scientist and person?
- How did my experiences as a mentor change my perception of mentorship?

8.1.2 Evaluate Yourself and Your Team

As an individual research mentor, you probably didn't (or couldn't) formally evaluate your performance as a mentor or the effective functioning of your research group (see Chap. 7). But you can informally evaluate yourself and your group. For example, start by reviewing the twelve most important elements of successful research teams (Box 4.1 in Chap. 4). What worked for you and your team, and what didn't? Did your undergraduate mentees contribute to making your research group stronger and more effective as a team? If they did, how did they do it? If not, why not? Remember, facing challenges as a team can make your team better and its research more successful.

But it's not enough to think about this in a vacuum. Figure out ways for your research group—including your undergraduate mentees—to let you know how they feel about group cohesiveness and doing scientific research together. These ways could include anonymous feedback (e.g., quick online or paper surveys) or discussions—with or without you in the room—facilitated by a trusted graduate student or post-doc in the research group. If you were part of a larger experiential undergraduate research program, there should have been an evaluation or assessment of the program (Chap. 7). Ask the program director how the program overall worked and how you did, what worked well, and how you might improve. Be open to constructive feedback and criticism.

8.1.3 Provide Feedback to Your Mentees

The last piece of stock-taking is to talk openly and honestly with your mentees about their experience working with you and your research group. Giving and receiving feedback can be difficult, but it is crucially important in your ongoing development as a mentor and in your mentees development as scientists. Your feedback to them and their feedback to you should reflect the joint expectations that you and your mentees set up at the beginning of their research experience (see Sect. 5.1 in Chap. 5).

In addition, you can rephrase the questions in Box 8.1 and ask your mentees how they think you did. Show them your own answers and look for commonalities and differences. Ask them: "How could I have done better?" By showing that you are open to feedback from them, you show that you trust them. This can help put them at ease and help them accept positive and negative feedback from you about their work.

8.2 Finishing up Your Mentees' Research Experience

You've reflected on your own work as a mentor, had a good evaluative session with your mentees, perhaps gotten additional feedback from your own colleagues, and maybe even taken a short vacation.[1] If you and your mentees have agreed that the experience was a positive one and you want to keep doing research together, what's next? At a minimum, you and your mentees know that a research project isn't really complete until you've presented your results to your STEMM peers and colleagues and archived your data so that others can use them in the future.

8.2.1 Tying up Loose Ends

There probably wasn't enough time for your mentees to "finish" the research in a way that you or they had hoped. Given the open-ended process that research is, did you really expect anything different? In fact, the lack of completion is an opportunity to continue mentoring your students in the process of doing STEMM research.

Chances are you and your mentees wrapped up some aspects of the project, but not others. They may have written an abstract of their results and prepared a poster describing it, but have they presented the poster to anyone other than the other people in your research group or undergraduate research program? Do they want to? (And, remember that it's their choice, not yours.) Maybe there are opportunities for them to present the poster at a group program for students in STEMM at your institution (or theirs, if it's different), or to colleagues at regional or (inter)national meetings. Perhaps you want to use the results of the project in a more expansive paper you could work on with your mentees or others in your research group. Alternatively, maybe there is the need to do some additional analysis or collect and analyze more data. Finally, it is essential that the data and any analytical code are well organized and digitally archived—for yourself, your colleagues, and for future mentees to use and build on.[2]

How much time would any or all of these activities take? A few weeks? Another semester? A year or more? Which piece(s) are most interesting to you and your mentees, which are most important to the broader scientific enterprise, and how much time are you and they willing to commit to doing them? The answers will vary among mentors and students, and will be strongly dependent on how much "ownership" your mentees feel about "their" project (see Sect. 5.2 in Chap. 5).

8.2.2 Data Management and Data Archiving

First things first. Learning to manage and archive data is the same as learning to write and organize information. Archived data and the accompanying documentation (metadata) is the 21st-century's lab notebook. Don't neglect this essential part of STEMM research!

If your mentees' data aren't clean and organized, they're not going to be able to analyze, interpret, and present them. So before you and your mentees even think about moving forward with presenting their research, deal with the data. Even if you think that your mentees' research data are incomplete or inadequate for future use, the importance of working with your mentees on data management cannot be overstated. Modeling your own commitment to open data and open science is at the heart of strong mentorship in STEMM. Take time to reiterate the issues of data "ownership" that you discussed with your mentees at the beginning of their research experience. Teach and learn with them how to organize, manage, clean, and archive data in ways that will make them usable by others now and in the future. Demonstrate through collaborative work that implementing "best practices" of data management[3] can take as much time as (or sometimes even more time than) doing the research itself.

8.2.3 Presentations at Meetings

One of the best way to keep students engaged after their research experience is to encourage them to present their research as a poster or an oral talk at a conference or meeting. We all want to share our findings with our communities and networks, and now our mentees are part of those same communities and networks. But teaching our mentees how to craft abstracts, posters, and talks (see Sect. 6.3 in Chap. 6) and encouraging them to present their work to others are only the first steps. They need real support, too. Financial support, of course (which of course, you budgeted for; see Sect. 2.2 in Chap. 2),[4] but professional and emotional support, too.

Think back on the first time you presented your own work at a meeting—as an undergraduate, graduate student, or post-doc. How did you feel? Did your mentor or advisor cut you loose on your own, or did they show up at your presentation and introduce you to their colleagues and past students? Did you eat alone or did your mentor include you in invitations to join groups going out to dinner after a day of exciting talks? Don't expect your mentees to navigate their first meeting entirely on their own. If you can't accompany your mentee to the meeting where they're going to present their work, enlist another member of your research group to go with, and be there for, your mentees in your stead. Many professional societies now have undergraduate chapters, and whether or not you can accompany your mentee to a meeting, you should ensure that they know about, and are connected to, these groups.

8.2.4 Writing It up

The research scientist's creed is *publish or vanish*, and almost nothing is as exciting to an undergraduate STEMM researcher as seeing their name in print on a paper that describes or includes the results of their work. Unfortunately, it's rare that a research project done during a single, short undergraduate research experience results in a peer-reviewed journal article.[5] But if the data your mentees collected are used in a paper that you or others in your research group write sometime in the future, don't neglect to recognize the contributions of your mentees. Whether you recognize their contribution through co-authorship or acknowledgment, it should have been discussed at the beginning of their research experience. Ideally, you should have discussed, and your research group should use, the standard criteria for co-authorship embodied in the CRediT (Contributor Roles Taxonomy) system (McNutt et al., 2018). But if it you didn't or wasn't, better late than never.

8.3 Additional Research Opportunities with and for Your Mentees

After their research experience, your mentees should appreciate that doing research always raises many more questions than it provides answers. If their research experience has solidified their commitment to an education and career path in STEMM, your mentees are going to be hungry for more undergraduate research opportunities (we discuss post-graduate research opportunities in the final chapter of this book). What can you do to help them find and take full advantage of additional research opportunities? And what's in it for you?

8.3.1 More Undergraduate Research with You

Mentees who are at your institution can easily do more research with you and your research group. They could continue on with the research project they started in their research experience or they could go in a new direction of mutual interest. For many undergraduates, continuing their research with you is very often done under the aegis of an independent study project or a senior thesis. But don't assume that is the route your mentees want to take. Rather, and as with their now-previous research experience, you and your mentees need to work out details of time commitment, funding *versus* academic credit, expectations, etc.

If you can commit the time, effort, and necessary financial support, continuing research with your mentees can have many tangible benefits. For example, you and your mentees could collect enough data to support a grant proposal (see Vignette 2.1 in Chap. 2) or get the project to a point where it is publishable. You also could enlist your mentees, now that they are more experienced members of your research group, as peer-mentors themselves, empowering them further while helping you with bringing new researchers into your group and onto STEMM pathways. But if you can't commit to being there for your mentees, be honest about it, and if they want to continue doing research with others, help them find a way to do it.

You will likely find it difficult to maintain the same level of research activity with mentees from different institutions. Most commonly, if they want to continue doing research, your mentees will move on to other projects with mentors at their home institutions. If your mentees develop their research with you into an undergraduate thesis project at their home institution, you may be able to serve on their thesis committee, but don't volunteer for the role. Empower them to ask.

Undergraduate mentees who choose to pursue post-graduate research may look to you as a possible MSc or PhD advisor. Based on our own experience, we discourage you from taking them on as graduate students (see also Chap. 9). Most undergraduate researchers develop a relationship and set of interactions with their mentors that is specific to their particular undergraduate research experience and their cognitive-developmental stage (Kegan, 1982). These methods and modes of interactions usually persist long after your mentees have completed their undergraduate research experience with you, and these types of interactions usually are not conducive to graduate or post-doctoral work.

8.3.2   Undergraduate Research with Others at Your Own Institution or Elsewhere

If your mentees are at your institution, you could also encourage them to continue doing research with one or more of your colleagues.[6] Based on your experiences with your mentees, you should have a pretty good idea as to who among your colleagues would work well with your mentees and who wouldn't work well with them. Talk honestly with your mentees about other possible mentors and with your appropriate colleagues about what kind of support your mentees are likely to need. You could also facilitate a group introductory meeting to help get your mentees and colleagues to know each other and begin working together. But very quickly, you should step back and let the new team figure out how to work together effectively (see also Chap. 9).

You can similarly encourage and help your mentees find and develop research opportunities at their home or other institutions. You may know some of the researchers at your mentees' schools with whom you could connect them. If there aren't opportunities for your mentees to do research at their home institutions, you may know of researchers at schools, agencies, or private labs in their areas. You can also help connect them to other off-site research experiences. In all these cases, you should expect that your mentees will contact you for suggestions and letters of recommendation. Be prepared to provide either or both, to the extent of your knowledge and abilities. And be honest with your mentees about the strength of the recommendations you can write for them and how much (or little) weight they are likely to carry in their application materials.

8.4   What's Next for You?

As you've learned from experience (or from this book), mentoring undergraduate researchers is different from mentoring (or advising or supervising) graduate and postdoctoral researchers. Working with students is energizing and a positive experience with mentoring undergraduate researchers usually leaves you ready to work with one or more new mentees.

But take a deep breath first and think it through. There is no "right" number, time, or frequency of undergraduate mentees. Every senior researcher has different amounts of time, energy, and goals for mentoring. It's most important that you know your capabilities, willingness to stretch, and your limits. Knowing yourself is the first step towards sustaining a career of mentorship.

Just as you don't have to take on new graduate students or postdocs every year, you are usually under no obligation to take on new undergraduate mentees every summer, semester, or year. Indeed, if you are continuing to work with one or more mentees on longer-term research projects, it may be to your and their benefit that you focus your attention on them instead of bringing more students into your research group. Maintaining this kind of focus is one route to avoiding "burn-out" from rapidly starting over with one or more new undergraduate research mentees. At the same time, if you do have a more senior undergraduate working in your group, you can work with them to jointly mentor another undergraduate.

8.5   From Mentors to Program Directors: Facilitating and Sustaining Further Undergraduate Research

Undergraduate research programs provide additional opportunities and ways for students to continue their research beyond their initial research experience. Rapid improvements in, and implementation of, virtual undergraduate research programs not only are opening up new avenues for facilitating continuation of undergraduate research, but also are making these opportunities available for many more students than could be accommodated in on-site, in-person programs (Vignette 8.1).

8.5.1   Facilitating Interactions

Whether your undergraduate research program draws all its mentors and students from a single institution or multiple institutions, as the program director you can identify pairings or groupings of mentors and mentees that naturally lead to continuation of the mentored research after the program ends. For example, you can encourage student participants who are considering senior thesis research to work with mentors who have a good track record mentoring senior thesis projects. If you have mentors in your program who are at another institution, encourage them to identify students at their home institution who could benefit from the program and who would be interested in continuing their mentored research after the program ends.[7]

8.5.2   Incentives for Continuing Undergraduate Research

Most undergraduate research programs provide students with research experiences that range from a few weeks to a few months. What incentives can you as a program director offer your cadre of mentors to continue to work with their mentees?

Most senior researchers mentor undergraduates because they want to educate new STEMM researchers. You can help your mentors out by providing supplies, equipment, and even space to facilitate the mentees' research projects. This kind of support may be especially valuable when the mentor is between grants or the student's project is difficult to tie directly to (and fund off of) an existing grant. Undergraduate research program budgets often have more flexibility in types of projects they can support than a research PI's targeted grant. Your program also can provide training courses and workshops in professional skills so that individual researchers don't have to (see Chap. 6).

Similarly, undergraduate research programs often budget for student travel to regional or national scientific meetings. This can augment travel funds available in the mentor's grant so that they can attend the meeting with their mentee. Funding also may be available for an entire cohort of student participants to attend a meeting together with the program director.

Some undergraduate research programs also provide modest stipends or direct research support to mentors in exchange for their participation in the program.[8] For researchers on nine-month ("academic"-year) contracts who rely on grants for "summer" salary or for "soft-money" researchers who are partly or entirely dependent on grants for their own salary, a stipend for mentoring can be a strong incentive to continue working with their mentees.

8.5.3   Take Advantage of Online Platforms

Research is international and researchers in multiple locations routinely collaborate from across large distances. We have regular group meetings on Zoom, Teams, or WeChat, use "the cloud" to process and analyze our data, and produce and upload videos to be presented at conferences held in other time zones. There's no reason that our undergraduate mentees can't do the same.

Undergraduate research programs can facilitate and support interactions between mentors and mentees before, during, or after an on-site, in-person research experience. For example, the program may be able to provide its participants with access to online platforms at a scale that a single institution cannot afford. Given the vagaries and restrictions of funding and differences in students' abilities, undergraduate research programs done entirely online are emerging, too. These online programs have the potential to dramatically extend the geographic reach of undergraduate research opportunities and expand the number and diversity of the programs' mentors and mentees (Vignette 8.1).

Vignette 8.1   Sustaining undergraduate research programs by adapting to the times

Contributed by the 2021 Polymath Jr. Mentors[*]

The COVID-19 pandemic upset the status quo, postponing or cancelling numerous research opportunities for undergraduate students. In early 2020, we already knew the coming summer couldn't be like the previous summers. We explored opportunities to move to virtual research experiences. This led to the creation of the Polymath Jr. REU (now supported by the US NSF). It's a combination of a standard Polymath project and a summer REU. Research mentors propose problems and then give students tremendous freedom to explore and solve them. Each project has post-docs or graduate students as additional near-peer mentors.

In 2020 we had 350 applicants, and accepted every student with at least one theory class and one letter of recommendation. We accepted 300 students, as we did not have to provide student housing or pay stipends. It was a great success.

In 2021, we expanded Polymath Jr. with more mentors and students, creating an international program offering research experiences to hundreds of students who would otherwise not have had this opportunity. The challenges of the pandemic inspired us to rethink how we mentor, develop the next generation of mentors, and engage students through new modalities.

[*] *The 2021 Polymath Jr. Mentors were: Kira Adaricheva, Zhanar Berikkyzy, Benjamin Brubaker, Marion Campisi, Patrick Devlin, Johanna Franklin, Seoyoung Kim, Steven J. Miller, Victor Moll, Christopher O'Neill, Luis David Garcia Puente, Victor Reiner, Eric Rowland, Alexandra Seceleanu, Adam Sheffer, Zoran Šunić, Enrique Treviño, Ezra Waxman, Yunus Zeytuncu, and Alexander Zupan.*

8.6 Take-Home Messages

✔ Take time to reflect on your experience mentoring undergraduate researchers.
✔ If you do continue research with your mentees, do it in ways that are effective for you and for them.
✔ Pace yourself as a mentor to avoid burning out.

References

Kegan, R. (1982). *The evolving self: Problem and process in human development*. Cambridge: Harvard University Press.
McNutt, M. K., Bradford, M., Drazen, J. M., Hanson, B., Howard, B., Jamieson, K. H., Kiermer, V., Marcus, E., Pope, B. K., Schekman, R., Swaminathan, S., Stang, P. J., & Verma, I. M. (2018). Transparency in authors' contributions and responsibilities to promote integrity in scientific publication. *Proceedings of the National Academy of Sciences, 115*(11), 2557–2560.

Chapter 9
Fledging Your Mentees

Abstract In this final chapter, three important topics involved in guiding one's mentee in new directions are addressed: (1) the range of post-graduate opportunities for your mentee; (2) writing strong letters of recommendation (and when and how to tell your mentee that you can't write a strong letter); and (3) moving forward with mentoring for yourself and your student. The chapter concludes with a discussion of how to recognize success—in oneself as a mentor and in one's mentees.

Just as young birds need to leave the nest and fly, the time always comes for students to move on from their formal research experience. In this final chapter, we address important topics involved in guiding one's mentee in new directions, focusing on the range of post-graduate opportunities for your mentee, writing strong letters of recommendation (and when and how to tell your mentee that you can't write a strong letter), and moving forward with mentoring for yourself and your student. We conclude with a discussion of the concept of "success;" the one Vignette in this chapter (Vignette 9.1) illustrates an example of how to recognize success—in oneself as a mentor and in one's mentees.

9.1 Post-graduate Directions

By mentoring students through their undergraduate research experience, you have given them a better understanding of what research in STEMM actually is and what a career in it could look like. For your mentees, their undergraduate research experience is very likely to have been a spark illuminating a deeper interest in STEMM. Our hope as STEMM professionals is that this deeper interest leads your mentees to further research, graduate or professional school, and a STEMM career.

Because you have been successful in your STEMM career, you are in a good position to advise and support your mentees on the path that you followed. But it's also important to be open to discussing career paths that differ from yours. For example, if you're in an academic position, recognize that many opportunities for rewarding careers in STEMM research exist in government, industry, or through

A. M. Ellison and M. V. Patel, *Success in Mentoring Your Student Researchers*,
SpringerBriefs in Education,
https://doi.org/10.1007/978-3-031-06645-0_9

self-employment. If you aren't familiar with these, there's no better time to learn about them than right now while you're helping your mentees find people who are more familiar with a range of career paths.

Your mentees also may decide that they'd rather pursue a different path. We encourage you not to frame that decision as a "loss" to the field or a "leak" in the STEMM "pipeline" (see also Fig. 4.1 and Sect. 4.1 in Chap. 4). Rather, celebrate their choices and consider how much happier your mentees will be for making that decision now instead of 20 years in the future. And encourage your mentees to remain scientifically literate while pursuing their own goals.

9.1.1 Employment

If you ask your mentee what they're going to do after they finish their undergraduate degree, a common response is that they're going to "take time off" and get a job before …well, before what? What they're probably thinking about is that they've had enough schooling for now, but because you got an advanced degree and you are their mentor and role model, that they will eventually follow in your footsteps and "go back to" school for another degree. You may have inadvertently encouraged this kind of thinking by referring to any career path other than yours as a road to an "alternative" career—and by implication, a much less desirable or rewarding one.

In fact, most graduates of two- and four-year colleges enter the workforce and don't pursue advanced graduate or professional degrees. For example, in the USA, among the more than 15 million college graduates <76 years old and with at least a BA or BSc in a STEMM or STEMM-related field, only 29% also had a MA or MSc 10% had a professional degree, and 7% had a PhD (Milan, 2019).[1]

As a mentor, you should encourage your mentees to think of a job in STEMM as the next step on their continuing career path. Emphasize how much their research experiences with you—and any products derived from it—can give them a competitive edge in getting a job doing research in STEMM. If they are vague or unfocused in their long-term career goals, suggest that they consider doing an internship to build their skill-set and get additional experience. Use your networks and connections to identify good opportunities for internships or jobs. And remind them that as a STEMM researcher in any job, they will always be learning more.

9.1.2 Post-baccalaureate Programs

If you were actively recruiting students for their potential (see Chap. 3), your mentees may need additional coursework before they'll be competitive for the STEMM jobs they're interested in or have the necessary prerequisites for graduate or professional schools. A number of colleges and universities offer one- or two-year post-baccalaureate ("post-bacc") certificate programs in STEMM fields. It is important to note that, as a rule, post-bacc programs are course- or skills-based and lead to a certificate or assertion of qualifications, not an additional degree.

In the US, post-bacc programs can be found in all fields, including all STEMM fields, but most of these programs are geared towards students interested in biomedical careers.[2] Significant federal funding is available to support diverse students who need further coursework or lab experience in the field.[3]

In Europe and the UK, post-graduate "conversion" and vocational courses, certificates, and diplomas are offered by many schools to students who are looking to learn skills or take classes in a field other than that in which they received their undergraduate degree.[4]

9.1.3 Graduate and Professional Schools

For many undergraduates, a research experience is also their initial introduction to the possibility of advanced study. The stated goal of most agencies and foundations that support undergraduate research is to encourage the students they support to enter careers in STEMM. Their most common measure of "success" for individuals mentoring undergraduates in STEMM research is how many of their mentees pursue graduate or professional degrees (i.e., MSc, PhD, or MD/PhD). And one of the most important outcomes of recruiting diverse undergraduates into research is the further diversification of graduate and professional programs and university faculties. If your mentees discover or strengthen an interest in pursuing an advanced degree in STEMM, you should do everything you can to encourage this interest and support their applications.

Many undergraduate research mentors and programs direct their mentees into graduate programs at the same institution. Because you have already invested substantial time and effort working with your mentees, it can be tempting to encourage them to continue on as graduate students in your research group. We discourage that path. As we noted in Chap. 8, mentees develop a relationship and set of interactions with their mentors that is specific to a particular time, research environment, and their cognitive-developmental stage (Kegan, 1982). These methods and modes of interactions usually persist even after mentees have completed their degree, and usually are not the set of behaviors that graduate advisors or mentors look for in their graduate students.

As a mentor, you have provided your undergraduate mentees with a solid foundation in STEMM research. Your continued growth as a mentor depends on your moving forward, learning from the experience, fledgling the mentees, and taking on new ones. Get ready for new challenges and you will be pleasantly surprised how soon your mentees will be your friends and colleagues (Vignette 9.1).

9.2   Letters of Recommendation

You will be regularly be asked to write letters of recommendation for your mentees as they apply for internships, permanent jobs, and applications for fellowships, and admission to graduate or professional schools. You know from experience what kinds of letters you expect to read when considering applications to your research group or department (Sect. 3.2 in Chap. 3). These are the same kinds of letters you should write if you want your mentees to succeed in STEMM,

Your assessment of the work you did with your mentees, the feedback sessions you had with them, and your own self-reflection (Sect. 8.1 in Chap. 8) should allow you to write a comprehensive letter tailored to the specific application and that focuses on the individual mentee's strengths (and weaknesses) and their potential for further successful research.

Your mentees need you to write not only a comprehensive letter, but also a strong one. If you don't think you can honestly write strong letters for your mentees, now is the time to have that discussion with them. We also note that as their mentor, there will come a time when you are no longer in a good position to write letters of recommendation. Although the length of time will differ among your mentees, as they move forward in their careers, their networks will expand and there will be others who are able to comment more effectively on their current work. You need to not only recognize when you're less knowledgeable about their activities, but also to communicate that effectively to your mentees.

9.3   From Mentees to Mentors

Mentorship is one of the most important skills you can help your mentees learn and develop. You should want your mentees to be your successors—not in your job or your career, but in the ways they can support future members of the STEMM community.

As we have illustrated throughout this book, the specific actions and behaviors that contribute to being a successful mentor can be described, taught, learned, and practiced. But in our experience, the best way to teach mentorship is to be a good mentor yourself. Walk the walk, don't just talk about it (Vignette 9.1).

Vignette 9.1   Long-term rewards of mentoring

Contributed by Sydne Record and Aaron M. Ellison

Sydne: As the first person in my family to pursue a doctoral degree, I didn't know how to get an advisor for graduate studies. As a result, I found myself in a graduate program without an advisor, but with an option to do lab rotations to find one. In a major stroke of luck, I met Aaron during a field course before my first semester, and he took me on as a student and mentee. He has fostered my professional growth at every step from graduate student to faculty member, from mentee to mentor, and as I've taken on multiple leadership roles. Aaron's mentorship has supported me throughout my career. Now, I'm paying it forward to the next generation of mentees and mentors as the new co-Director of the Harvard Forest Summer Research Program [HF-SRP].

Aaron: Over four decades,I have had the privilege of working with hundreds of students—undergraduates, graduate students, post-docs—in relationships ranging from managerial to mentoring (Fig. 1.1 in Chap. 1). When I first met Sydne, and then as her dissertation advisor, my goal was to support her through graduate school and set her on course for her professed goal of an academic career. But the rewards of mentorship go far beyond that goal. While she was a graduate student, Sydne started mentoring students in the HF-SRP and continued to do so as she progressed in her career. To my pleasant surprise, when I was stepping down as co-Director of HF-SRP, Sydne unexpectedly asked to be considered for the role, and in 2020 stepped into the position. Having played a part in my mentee becoming a mentor of mentors is my greatest long-term reward of mentorship.[5]

9.4   From Mentors to Program Directors: Finding Your Successors

Anyone who has developed and run an experiential undergraduate research program in STEMM knows that creating and maintaining it requires much more than the science skills you use every day as a research mentor. You also need administrative and managerial skills that researchers are rarely taught. And then there is a range of other less tangible "people" or "soft" skills that we may not even recognize we're using. These skills are essential for building and cultivating the strong and energetic teams of research mentors and administrative support staff that keep the whole operation running effectively and efficiently.

We also know that no one person should run an experiential undergraduate research program "forever." Because most such programs are funded by grants, program directors are constantly working to secure new or renewed funding for them. Programs evolve in tandem with changes in scientific fields and with changing approaches to pedagogy and mentorship. It can be hard to maintain your enthusiasm and avoid burning out. In our own experience, 10–15 years is a reasonable amount of time to to develop an idea or structure for the program, work out its implementation, get it funded, direct it, and evaluate and iteratively improve it. And perhaps most importantly, if you think the program should outlast your tenure running it, you should work to effect a smooth transition in the program leadership.

We do not suggest identifying a particular person as your successor. All of us are more likely to pick someone a lot like ourselves—someone who shares our disciplinary focus and our way of thinking about mentoring undergraduate research. Rather, we encourage you to model good leadership skills and behaviors and offer to mentor any or all of your research mentors in the skills and practices of being a leader. Not all of your mentors will be interested in the opportunity to learn the skills they'll need to direct an undergraduate research program. But some will. When the time comes to pass the baton, there will be many who will volunteer to take it on and move it forward.

9.5   Take-Home Messages

✔ Mentorship is for life, but the level of support you give your mentees will change as they fledge.
✔ The best way to teach mentorship is to be a good mentor.
✔ Success comes in many forms. Celebrate all of them.

References

Kegan, R. (1982). *The evolving self: Problem and process in human development*. Cambridge: Harvard University Press.
Milan, L. (2019). *Characteristics of college graduates, with a focus on veterans*. Retrieved December 02, 2021, from https://www.nsf.gov/statistics/2019/nsf19300/.

Notes

Notes for Chap. 2

1. These budget categories correspond to those seen in US NSF, NIH, DOE, and USDA budget sheets, but similar categories are used in grant proposals submitted throughout the European Union or to private philanthropies worldwide. You should also note that costs for undergraduate student researchers may be in categories that are subject to indirect costs ("overhead") or in "Participant Support Costs," which are not subject to indirect costs. If you are writing a proposal that includes support for undergraduate researchers, it is a good idea to check with your sponsored research office or the program officer at the granting agency to determine what is allowable in budgeting these costs.
2. In our experience, students and more senior researchers rarely understand the distinction between a salary and a stipend. A salary is paid regularly (usually weekly or biweekly) to employees based on work done (e.g., as an hourly wage) and is taxed. A stipend is usually paid in larger blocks (monthly or for the entire project) to interns or trainees and may or may not be taxed. Payment of a stipend implies a mentor-mentee relationship, whereas salaries are expected in an employer-employee (i.e., managerial) relationship (Chap. 1, Fig. 1.1). Stipends are fixed and independent of performance, whereas salaries may change through time based on performance. Although you might expect that all undergraduate researchers receive stipends, some institutions treat paid undergraduate student researchers as temporary employees who receive a salary so that they can be covered by, for example, workers' compensation insurance.
3. A minimum wage is the minimum salary (or hourly rate) set by law in a country or other jurisdiction. A living wage reflects the actual cost of living in a particular area, and is usually significantly higher than the minimum wage. Examples are easily available on the web, for example for the US or the UK.
4. For example, undergraduates receiving financial aid from their college or university may be expected to earn and contribute a certain amount each year to their tuition, room, and board costs. If those required earnings are just (or not quite)

© The Author(s), under exclusive license to Springer Nature Switzerland AG 2022   93
A. M. Ellison and M. V. Patel, *Success in Mentoring Your Student Researchers*,
SpringerBriefs in Education,
https://doi.org/10.1007/978-3-031-06645-0

covered by the stipend you are paying, there won't be anything left for rent and food.

5. To save these costs, you might ask your student to stay on campus after the term ends and before starting their research with you, or stay on campus after the research experience has ended but the next term hasn't started yet. But then, you should budget for those additional subsistence costs.

6. Meeting travel costs for undergraduates are no different from those for more senior researchers: registration, transportation, housing, and food (or *per diem*).

7. For most grants supported by US federal agencies, stipends, subsistence, and travel included as participant support are not subject to indirect costs (overhead). For students receiving financial aid through their college or university and eligible for work-study support, federal work-study funds (outside of your research grant) normally cover 50–75% of the student salary or stipend.

8. Shannon Ladeau, the undergraduate lead of this project, went on to do her PhD at Duke University in Jim Clark's research group. She is currently a Disease Ecologist at the Cary Institute of Ecosystem Studies, where she also mentors undergraduates engaged in research in the Cary Institute's long-running NSF REU Site.

9. Francesca Meier braved the turtles while working with the sponges. She completed her BA in Maritime Studies and Philosophy and went on to study massage and nursing.

10. Sybil Gotsch, the undergraduate who helped Aaron launch this work, completed her senior thesis on the germination ecology of the northern pitcher plant *Sarracenia purpurea*. She went on to complete a PhD in ecology with Manuel Lerdau at SUNY Stony Brook. She is currently an Associate Professor at Franklin & Marshall College, where she has mentored many undergraduate researchers.

11. The data collected by undergraduates in the summer of 1995 provided crucial preliminary data for an NSF proposal submitted a year later. This proposal was the first in a series of grants on the pitcher-plant system awarded to Aaron and his long-time friend and collaborator, Nick Gotelli, and that supported dozens of undergraduate research students from 1998 to 2015.

12. For more on this, see Chap. 4 in our companion book, *Success in Navigating Your Student Research Experience*.

13. Indeed, the America COMPETES Reauthorization Act of 2010, US Public Law 111–358, 42 USC 1861 requires in §514 that "mentors and students are supported with appropriate salary or stipends" at REU Sites supported by NSF (42 USC 1862p-6). In our experience, however, adding such stipends to otherwise tight NSF REU Site budgets is rarely possible.

14. For example, NSF does not allow undergraduate research stipends to be split among two or more REU Site awards.

Notes for Chap. 3

1. Definitions from the US National Institutes of Health's Implicit Bias Training Module.

2. When inviting yourself to give a recruiting talk, you should also indicate that you are not asking for any support from the institution or organization you are visiting. You should do this on your own dime or have built it into your research grant's travel or recruitment budget, arrange your own transportation and, if necessary, any meals and accommodation.

3. For additional advice to students on this topic, see Chap. 2 in our companion book, *Success in Navigating your Student Research Experience.*

4. Keep in mind that obtaining formal transcripts may cost the student money that they don't have, reducing your applicant pool in unanticipated ways.

5. A good platform for identifying good meeting times in different time zones is http://www.worldchatclock.com/.

6. If the research opportunity actually requires the use of videoconferencing technology, that should have been specified in the position description and a student requesting a phone, not video-conferenced, interview may be a "red flag."

7. If you don't know what these are, ask someone from your Human Resources department.

8. At the same time, many organizations that support undergraduate research programs prioritize applications from groups whose mentors have their own research funding that can supplement or cover completely the day-to-day costs of the students' lab or field work.

Notes for Chap. 4

1. In fact, the name of the type of graph shown in Fig. 4.1—an *alluvial diagram*— refers to the distribution and deposition of materials by flowing water. Although originally developed as a way of illustrating temporal changes (analogous to a pipeline), it can also be used, as we do here, for summarizing demographic or synthetic information.

Notes for Chap. 5

1. We discuss student expectations and how they can best express them to you in Chaps. 1 and 4 of our companion book, *Success in Navigating your Student Research Experience.*

2. For more on this topic, see Chap. 1 in our companion book, *Success in Navigating your Student Research Experience.*

3. See the Vignette 4.1 in our companion book, *Success in Navigating your Student Research Experience* to learn more about why not to tie one's sense of self-worth and success in the research experience to statistical "significance" and *P* values.

Notes for Chap. 6

1. The definition of responsible conduct of research (RCR) that we use here is adapted from an overview of RCR written by the US National Institutes of Health Office of Intramural Research.

2. Good sources for USA guidelines on RCR include the US National Institutes of Health Office of Intramural Research, the US National Science Foundation

Office of the Director, the American Psychological Association, and the European Network of Research Ethics and Research Integrity.

3. In the USA, these topics will also be covered in additional training courses and workshops focused on compliance with "Title IX." Title IX is a federal civil rights law (Public Law No. 92–318, 86 Stat. 235; codified as 20 U.S.C. §§1681–1688) passed in 1972 that prohibits sex-based discrimination in any school or other education program that receives federal money.

4. The FAIR guidelines—making sure your data are Findable, Accessible, Interoperable, and Reusable—encapsulate current best practices for responsible and ethical data management.

5. The CRediT system provides clear guidelines for maintaining integrity and transparency in determining and asserting (co-)authorship of research papers. Open-access refers to Creative Commons licensing.

6. We discuss how students can develop themselves as mentors in Chap. 9 of our companion book, *Success in Navigating your Student Research Experience*.

Notes for Chap. 7

1. For example, Since 2005, the US National Science Foundation (NSF) has emphasized the use of project evaluations to measure, both qualitatively and quantitatively, the success of REU programs. Participant tracking for at least seven years following participation in an undergraduate research experience, including collecting data on graduation from college, matriculation in STEM graduate programs, and entry into the STEMM workforce has been required for undergraduate research experiences supported by the all NSF directorates since the passage and implementation of the America COMPETES Act of 2010 (42 USC 6621: Coordination of Federal STEM Education).

2. Primary reasons for restricting formal evaluations and surveys to cohorts include the desire for sample sizes sufficient for qualitative or quantitative analyses and the ability to maintain anonymity of survey respondents. If you are doing a formal evaluation of either a program or your own research group, and especially if you are working with a professional evaluator or education researcher, your survey instruments and protocols should be reviewed and approved by an Institutional Review Board (IRB), which oversees research on human subjects. If you think you will ever share your data with others or publish your data, your evaluation protocols *must* be approved by the IRB. On the other hand, IRB approval is not required for less formal evaluations of you and your own research group.

3. Although these self-assessments are often characterized as "summative" evaluations, they usually lack a baseline, comparable assessment of initial conditions (i.e., before the research experience started) or other comparative benchmarks.

4. There are few comparable examples of formal undergraduate research programs in countries outside of the US. A notable exception is the Undergraduate Research Experience Program of the Qatar National Research Fund.

5. Although HHMI is a biomedical-based research institute, for many years it funded research programs in all STEMM fields at undergraduate institutions in the US.

6. Only two genders—male and female—were coded in the 2003 SURE survey. Ethnicities coded in the 2003 SURE survey were African-American, Asian-American, Caucasian, Foreign National, Hispanic, Native American, Other, and Multiracial.

7. URSSA is based on the widely-used Student Assessment of Learning Gains (SALG). Hunter (2009) noted that URSSA was appropriately used only for undergraduate research programs with > 10 students (to preserve anonymity). URSSA could be given over the course of an undergraduate research program (e.g., before, during, after) to formatively gauge student improvement through time, but in practice, it was normally given to students only at the end of their research experience, providing a summative evaluation.

8. Unlike URSSA, which was made available to the research community at no cost, CIMER charges several thousand US dollars to set up and support a survey "project" for each program that uses it.

9. The Harvard Forest Summer Research Program in Ecology is an REU site that has been supported by the US NSF and other federal agencies since 1992.

10. The surveys and associated research were approved by the Harvard University Institutional Review Board, protocol numbers F14015-102 [2007], F16874-102 [2009], 14-2580 [2014], and 15-2719 [2015].

11. In addition to its comprehensive platform and instruments for evaluating students who participate in undergraduate research programs, CIMER also has tools to evaluate mentors.

Notes for Chap. 8

1. After decades of mentoring undergraduate researchers during summers and winter intersessions, we have found that there's nothing like a 2-week vacation to fully recharge before resuming our own research and mentorship activities.

2. We emphasize that data should be archived in a public archive that is "permanent" (i.e., has a reasonable likelihood of outlasting your career), that will assign a persist digital object identifier (DOI) to the data or code, and that is indexed and findable by search engines. Examples familiar to scientists include Dryad, Figshare, GenBank, EDI, and the wide range of databases used for crystal structure and data. There are many others, but your own personal lab website, your departmental or institutional website, and GitHub are not considered permanent repositories.

3. Best practices for data management start with the FAIR principles for scientific data management and stewardship.

4. Consider your own past experiences. Did your mentor cover your travel, housing, and meal costs, did they ask you to max out your credit card and get reimbursed later, or did they expect you to pay out of your own pocket? Only the first option is reasonable and equitable, especially for undergraduate researchers (see Chap. 2).

5. There are several peer-reviewed journals dedicated to publishing undergraduate research. Two well-established ones are the *American Journal of Undergraduate Research*, published since 2002, and the *Journal of Undergraduate Research and Scholarly Excellence*, published since 2010.

6. We do not include your graduate students or post-docs in the "your colleagues" category. Although it is reasonable to ask or encourage your graduate students or post-docs to be the primary mentor for your mentees after they've completed their research experience, if your graduate students or post-docs agree to take this on, you will still have to be ready to continue to be a mentor and be there for your mentees when the need arises. So, if you couldn't commit the time, energy, or resources to doing more mentored research with your mentees, it's usually not a good idea to ask your graduate students or post-docs to pick up your slack.

7. If your program budget doesn't include participation by mentors and mentees from other institutions, consider developing an MOU [Memorandum of Understanding] or cost-share agreement with the other institution(s) to support such participation. These MOUs and other agreements can expand the size and diversity of your program, too.

8. Indeed, §514 of the America COMPETES Reauthorization Act of 2010 (US Public Law 111–358) requires that grants awarded by the US NSF to support Research Experiences for Undergraduates include funds to ensure that "mentors and students are supported with appropriate salary or stipends."

Notes for Chap. 9

1. These data are from the 2017 National Survey of College Graduates conducted by the US National Science Foundation. In her 2019 article, Milan presents a more fine-grained analysis of these data.

2. The Association of American Medical Colleges maintains a database of available post-bacc programs and opportunities.

3. The goal of the US NIH's Postbaccalaureate Research Education Program (PREP) is to "develop a diverse pool of well-trained postbaccalaureates who will transition into and complete rigorous biomedical, research-focused doctoral degree programs (e.g., Ph.D. or M.D./Ph.D.) in biomedical fields." Support for PREP funds is limited to major research universities (R1s) that already have education and training programs and grant Ph.D.'s in core biomedical fields.

4. For example, the British Council maintains listings for non-degree post-graduate opportunities in the UK. A broader online resource is at postgrad.com.

5. An additional reward has been publishing papers with Sydne and her students, which makes me a grand-mentor!